Robert Greenberg

Real Existence, Ideal Necessity

Kantstudien

Ergänzungshefte

im Auftrage der Kant-Gesellschaft
herausgegeben von
Manfred Baum, Bernd Dörflinger
und Thomas M. Seebohm

157

Walter de Gruyter · Berlin · New York

Robert Greenberg

Real Existence, Ideal Necessity

Kant's Compromise, and the Modalities without the Compromise

Walter de Gruyter · Berlin · New York

♾ Printed on acid-free paper which falls within the guidelines
of the ANSI to ensure permanence and durability.

ISSN 0340-6059
ISBN 978-3-11-020690-6

Library of Congress Cataloging-in-Publication Data

A CIP catalogue record for this book is available from the Library of Congress.

Bibliographic information published by the Deutsche Nationalbibliothek

The Deutsche Nationalbibliothek lists this publication in the Deutsche Nationalbibliografie;
detailed bibliographic data are available in the Internet at http://dnb.d-nb.de.

Printed in Germany
Cover design: Christopher Schneider, Berlin
Printing and binding: Hubert & Co. GmbH & Co. KG, Göttingen

To Claire, Hank, Sasha, Maggie, Leo, and Liza,
and the joys of familiarity and discovery

Acknowledgements

Earlier versions of portions of this book have been read at various philosophical conferences and meetings, and among these, some have been published. An earlier version of much of *chapter 3* was read at the American Philosophical Association Meetings, Eastern Division, 2004. I wish to express my thanks to Tim Rosenkoetter for his useful comments on the paper. And I want to thank the editor and the publisher of *Kantian Review* for their kind permission to use some of that material in an article that appeared in *KR*, vol. 11, 2006. The same *KR* article also contains much of a paper delivered to the American Philosophical Association Meetings, Central Division, 2004, and I am indebted to Gary Hatfield for his helpful comments on that paper, which is also an ancestor of a portion of a different chapter in this book, *chapter 9*; and I would like to thank the editor and the publisher of *KR* for permission to reprint that material. *Chapter 4* contains material that appeared earlier as a paper delivered at the American Philosophical Association Meetings, Central Division, 2002, and I would like to thank Gordon Brittan for his comments on that paper. A subsequent version of the paper was delivered at the 21st World Congress of Philosophy, 2003, Istanbul. *Chapter 5* originated as a paper delivered at a symposium shared with Robert Hanna and Robert Howell, commentator, on Kant and Analytic Philosophy, at the Meetings of the American Philosophical Association, Eastern Division, 2002, and I would like to thank Howell for his comments on my paper. An earlier version of a portion of *chapter 8* was delivered at the 10th International Kant Congress, Sao Paulo, 2005, and published in the *Proceedings* of the Congress. I would like to thank the editors and the publisher of the *Proceedings* for their kind permission to include material from the paper in this book.

I am also indebted to Kenneth Edward Walden, a Kant student knowledgeable in modal logic, who read through an earlier version of the manuscript with an eye to the adequacy of my uses of modal logic, and in particular, the logical formulations and the derivations of the modalities in *chapters 10* and *11*. Finally, I would again like to thank Richard Aquila for his continuing encouragement of and assistance in the publication of my work on Kant generally.

Inhalt

Preface

My previous book on Kant, *Kant's Theory of A Priori Knowledge*,[1] henceforth abbreviated as *Kant's Theory*, focused on a certain *logical* conception of the ontology of Kant's theory – a conception adapted from Quine's notion of ontology. The present book instead focuses on a certain *epistemological* conception of the ontology of Kant's theory of our *a priori* knowledge of objects – a conception derived from Kant's own description of the beginning of our knowledge of existing objects. But, like the previous book, the primary interest is in Kant's *theory* of knowledge – his epistemology – and not the knowledge whose possibility the theory is supposed to explain. In line with my interpretation of Kant's description of the beginning of our knowledge of existing objects, an interpretation that is given in chapter 2, the objects involved in this beginning are identified as *the external objects that get our knowledge started*, which, because they start our knowledge, will henceforth be called simply, ***initiators.***[2] Since they constitute the ontology of Kant's theory of *a priori* knowledge as seen from an epistemological point of view, the present book offers an epistemological interpretation of the ontology of Kant's theory. The use of the definite article in both cases – the logical and the epistemological conceptions of *the* ontology of the theory – is appropriate, since the theory is actually committed to just one ontology, previously conceived logically, but now conceived epistemologically. The following paragraphs describe the change in focus between the two conceptions of Kant's ontology.

In *Kant's Theory*, it was argued that the objects of the ontology constitute his basic reality, which I went on to interpret as objects of *transcendental affirmation.* Following Kant's own use of the term "transcendental," it meant that the objects belonged to Kant's *theory* of the possibility

1 *Kant's Theory of A Priori Knowledge* (University Park: Penn State Press, 2001), p. 39. The term 'knowledge' in this title and in the book itself as well as in the present book is a translation of Kant's use of *'Erkenntnis'*, unless otherwise noted.

2 Thanks to Jonathan Greenberg for suggesting the name *"initiators"* for these objects.

of our *a priori* knowledge of objects – his *transcendental epistemology.*[3] In this sense of "transcendental," though space, time, and the categories are *a priori* cognitions for Kant, they are not transcendental, since they belong to the knowledge that the epistemology is supposed to explain, and not to the epistemology itself (A56/B89–81).[4] And the term "affirmation" meant Kant's eponymous logical function of judgment as found in his Table of Judgments.[5] Finally, the objects of transcendental affirmation were identified with Kant's "*things*" (*Dinge*).[6] This was based on his use of "*Dinge*" throughout his account of the possibility of *a priori* knowledge and that conformed to the meaning of "transcendental affirmation." It was also supported by his remarks about "being" and "thinghood" in "The Ideal of Pure Reason."[7] In *Kant's Theory,* therefore, basic reality was said to consists of Kant's *things* – these were the objects that Kant's epistemology posited as existing independently of us, and hence independently of our knowledge of such reality, to the extent that the knowledge itself involved something about us that is reciprocally independent of reality.[8]

As already noted, this conception of Kant's ontology was based on W. V. Quine's logical notion of ontology. In *Kant's Theory* I said, "By 'ontology' I mean, à la Quine, all the objects that must exist for all the af-

3 Ibid, p. 11.
4 Kant, *Critique of Pure Reason.* References to the *Critique* in the body of the text (A/B) are made in the standard way by using "A" and "B" to refer to the first and the second editions of the work and numerals for the paginations of the two editions. All quotations from the *Critique* will be taken from the Norman Kemp Smith translation, unless otherwise noted. The other sources will be the Werner S. Pluhar translation (Indianapolis/Cambridge: Hackett Publishing Company, 1996) and the Paul Guyer and Allen W. Wood translation (New York: Cambridge University Press, 1998)]. The particular translation of a passage from the *Critique* that appears in the text will be chosen on the basis of my own examination of the German text, edited by Raymund Schmidt (Hamburg: Felix Meiner Verlag, 1990), and an accommodation to English usage.
5 *Kant's Theory,* pp. 34–36.
6 Ibid, see the Index in *Kant's Theory* under "thing."
7 Ibid. pp. 34–35.
8 It is noteworthy that only after the publication of *Kant's Theory* did some of the foremost commentators on Kant come to adopt the terminology of 'thing' – instead of 'empirical object' or 'appearance,' which used to be the expressions of choice – when speaking of the independent, existing objects that Kant says we can know. See, for example, Allen Wood, Paul Guyer, and Henry E. Allison, "Debating Allison on Transcendental Idealism," *Kantian Review,* vol. 12-2, 2007. Ibid. pp. 34–35.

firmations of a theory to be true."[9] There are points of both comparison and contrast between the Kantian criterion of ontological commitment as put forth in *Kant's Theory* and Quine's criterion. First, the comparison: both criteria are *logical* in nature. Affirmation is a logical notion, pertaining to something's being said affirmatively of something else, instead of saying it, say, negatively. Kant calls it a *logical function of judgment.* The contrast between the two criteria, however, is that Quine's is a logic *without* an ontology, whereas the logic to which the logical functions of judgment in the *Critique* belong has its own ontology. Kant calls it *transcendental logic.* He would consider Quine's logic *general logic*, meaning that it abstracts from all content of judgment. Its terms, therefore, have no reference to any object whatsoever. The *determination* of an ontology for Quine is left to the *non-logical predicates* of sentences that are framed in terms of the logical apparatus of reference. Consequently, Quine's criterion of ontological commitment is totally neutral with respect to which objects belong to a theory's ontology.

Kant's transcendental logic, on the other hand, is the logic of a special knowledge, viz., our *a priori* knowledge of objects. It excludes our *empirical* knowledge of objects (A55/B79 ff.). This very constraint on the application of the logic gives it content, and hence possible reference to objects. On the basis of (most of) the logical functions of judgment,[10] Kant derives concepts that distinguish classes of objects from one another. *Reality* is such a concept, which is based on the logical function of *affirmation. Real* objects can accordingly be distinguished from other objects.[11] Consequently, affirmation can be the basis for distinguishing a certain class of objects independently of any *predicates* that may be ascribed to the objects in an affirmative proposition. Hence, Kant's criterion of ontological commitment is not ontologically neutral. Though the criterion remains logical in nature, it is not sufficiently abstract to be free of its giving grounds for a commitment to a particular ontology, namely, an ontology of *things.*

The new book – the one in hand – employs a criterion of ontological commitment that is at once both closer to and farther from Quine's cri-

9 Ibid. p. 18. See, W. V. Quine, "On What There Is," and "Logic and the Reification of Universals," in *From a Logical Point of View* (Cambridge, MA: Harvard University Press, 1980 [first published 1953]).

10 The reason for the qualification "most" is discussed in *Kant's Theory*, Chapter 9. The simple point there is that the modalities do not have content.

11 See "The Ideal of Pure Reason" in the *Critique.*

terion. It is closer in that it completely leaves the determination of the ontology of a theory to the *predicates* of the true affirmations of the theory. Consequently, affirmation itself is no longer used as the basis of an interpretation of Kant's ontology. But the criterion departs from Quine's in that the predicates in question belong to an *epistemology* – Kant's epistemology of *a priori* knowledge, whereas for Quine they belong to whatever knowledge or theory one is using to talk about existing objects, such as ordinary language or a particular science, such as physics or biology, but not an epistemology for talking about objects in general, such as might be found in Quine's *epistemology naturalized.*[12]

This particular departure from Quine is also a departure from the earlier interpretation of Kant's criterion of ontological commitment in the following respect. Whereas the earlier interpretation, framed as it was by Quine's notion of ontology, attributed to Kant a *logical* criterion of ontological commitment, albeit a *transcendentally* logical criterion, the present interpretation attributes to Kant an *epistemological* criterion of ontological commitment. And since the particular knowledge of which it is an epistemology is *a priori* knowledge, the epistemology is *transcendental.* So, in the present volume, the constituents of Kant's ontology will be determined according to certain concepts of Kant's transcendental epistemology. These concepts cover such notions of Kant's as our being affected by objects, the *a priori* modes in which we can be affected by them – the *a priori* modes of our intuition (which Kant claims to be space and time) – as well as the objects which can so affect us in such modes as to produce in us – a causal efficacy – what Kant calls *appearances.* These latter objects are Kant's things in themselves.

This introduces the topic of the correct interpretation of Kant's transcendental idealism – Kant's distinction between appearances and things in themselves and his claim that we can have *a priori* knowledge only of appearances, but not of things in themselves. It was argued in *Kant's Theory* that appearances and things in themselves are the very same objects – Kant's *things* – only determined in two mutually exclusive ways. One way depends on the *a priori* modes of our intuition and the other, on thought alone, or, more precisely, on the thought of objects *as being independent of any a priori modes of intuition,* and hence the thought of objects as *being merely thought.*

---

12 W. V. Quine, "Epistemology Naturalized," in *Ontological Relativity and Other Essays* (New York: Columbia University Press, 1969).

This puts the present interpretation of Kant's transcendental idealism in the camp of the so-called dual-aspect interpreters, as distinguished from the dual-object theorists, who hold that Kant was ontologically committed to two distinct types of object – appearance and thing in itself – with no single type of object spanning the two types in some way – a way in which appearance and thing in itself can be considered as mere aspects of one and the same object.

In the present book, it is also maintained again that Kant's transcendental idealism should be interpreted as a distinction between two different ways of determining a single set of objects. So again there is argument for a dual-aspect instead of a double-object interpretation of the idealism. The difference between my two dual-aspect interpretations of the idealism – that expressed in *Kant's Theory* and that found in the present work – reflects the new epistemological method of determining the identical objects in contrast to the previous logical method of doing so. Since the single set of objects is now interpreted as the external objects that get our knowledge started, i. e. as initiators, we must ask, Which objects are these? What are their most salient characteristics that are germane to our present study? In chapter 2 I present an account in which initiators are to be understood as the objects that *in affecting us* (in our *a priori* modes of intuition) *are given to us.* Kant's idealism is then explained in terms of the two aspects of initiators. In affecting us they produce appearances in our mind, and hence must be able to be thought as *being independent* of our power to receive representations of them. This is to think of initiators as *things in themselves.* However, they are also understood as being given to us, and hence must be able to be thought in *relation to* our power to receive representations of them. This is to think of initiators as *being dependent on a mode of our a priori intuition*, which is to think of them as *appearances.* So, since initiators can both affect us and be given to us, they can be thought in terms of the two aspects in question: as independent of, on the one hand, and as relative to, on the other, our power to receive representations of them. Kant calls this power *sensibility.*

To sum up, the present work still subscribes to a dual-aspect interpretation of Kant's transcendental idealism, only now Kant's ontology is determined epistemologically, not logically, where the epistemology distinguishes between different modes of references to objects.

Although the interpretations of Kant's two criteria of ontological commitment – the logical and the epistemological – are logically distinct from each other, they are nonetheless extensionally equivalent. That is, the same objects are picked out by both of them. Kant's objects of transcendental affirmation – *things* – are the same objects that in affecting us in our *a priori* modes of intuition are given to us through those modes – *initiators*. Despite this extensional equivalence, I will use the new term – *initiator* – to refer to these identical objects in order to mark the epistemological standpoint from which the present book is written.

In light of the change in focus from a logical to an epistemological criterion of Kantian ontological commitment, after the subject of the book is introduced in the first chapter – Kant's acceding to the realistic demands of existence at the very moment he provides an idealistic account of necessity – what is called in the title of the book, his "compromise," the actual account of the compromise begins in chapter 2, with an analysis of his epistemological conception of the objects that get our knowledge started, i. e. initiators. In the very beginning of the Introduction to both editions of the *Critique* Kant asserts that the *knowledge* in which *all* our knowledge begins is *empirical* knowledge, that is, *experience*, and experience for Kant entails that we are affected by independent objects. So, the epistemological approach to the *Critique* that characterizes the book naturally begins the account of Kant's compromise with an examination of the nature of *the objects that get our knowledge started.*

While the slant of the book is overtly epistemological, it is still not without two uses of logic – the first of which is employed in the second use as well. Both uses are tied in their respective ways to certain issues in philosophy of language. The first use involves the distinction in logical relation between entailment and presupposition. Given a logical relation of antecedent and consequent between two propositions, entailment is a relation in which the falsehood of the consequent results in the falsehood of the antecedent, whereas according to presupposition it is a relation in which the falsehood of the consequent results in the antecedent's being neither true nor false. Moreover, if a proposition is meaningless unless it has a truth value (a position held by Russell, but not by Frege or Strawson), and the logical relation is presupposition (a relation employed by Russell, as well as by Frege and Strawson, as chapter 5 will make clear), the falsehood of the consequent will also have the result that the antecedent is meaningless.

This distinction between these two types of logical conditional is used to distinguish between the existence of Kant's *initiators in space and time* and the existence of *appearances in space and time.* It thereby yields a distinction between initiators and appearances that is independent of the epistemological distinction between them, provided the concepts of existence and of space and time are kept constant in the two cases. Whereas a proposition affirming the existence of initiators in space and time *entails* that we intuit initiators, one affirming the existence of appearances in space and time instead *presupposes* that we intuit them, i.e. initiators. The independence of the distinction between the two *logical* relations from the *epistemological* distinction between initiators and appearances allows this *logical* determination of the distinction between the two types of object that was just drawn to go on to reinforce the previously developed *epistemological* distinction between them.

The other use of logic occurs in the discussion of Kant's uses of possibility and necessity, and in particular, his uses of *de re* necessity. In this regard, notations and techniques from recent uses of modal logic in philosophy of language, in particular, those belonging to David Kaplan in an essay he wrote from a Fregean stand point before he adopted direct reference theory, are employed to present an interpretation of Kant's uses of these modal notions in his account of the possibility of our *a priori* knowledge of objects.[13] Gareth Evans and John McDowell provide convincing arguments against those who deny that a thought, or proposition, can be given a *de re* interpretation from a Fregean standpoint.[14] Evans' and McDowell's work thus supports Kaplan's attempt to provide just such an interpretation of our uses of *de re* necessity, if what they have to say about Frege on propositional attitudes can serve as a background for using Frege to provide an interpretation of our uses of the *de re* alethic modalities, and in particular, *de re* necessity. Since Kaplan's interpretation of our uses of *de re* necessity is composed from a Fregean standpoint, and

---

13 David Kaplan, "Quantifying In," in *Reference and Modality*, ed. Leonard Linsky, (London: Oxford University Press, 1971) pp. 112–44. Originally appeared in *Words and Objections: Essays on the Work of W. V. Quine*, ed. by D. Davidson and J. Hintikka (Dordrecht-Holland: D. Reidel Publishing Co., 1969), pp. 178–214. His subsequent adoption of direct reference theory is found in his "*Demonstratives*," in *Themes from Kaplan*, eds. Joseph Almog, John Perry, and Howard Wettsetein, (New York, 1989), pp. 481–614. His revised notions of necessity can be found especially in pp. 593–99. Only Kaplan's interpretation of *de re* necessity as found in "Quantifying In" will be used as a model for our own interpretation of Kant's uses of *de re* necessity in his interpretation of our own uses of the necessity in our thought and discourse.

14 Gareth Evans, *The Varieties of Reference*, ed. by John McDowell (Oxford: Clarendon Press, 1982), chaps. 1, 6, and 7, cited by John McDowell, *Meaning, Knowledge, and Reality* (Cambridge, MA: Harvard University Press, 1998) p. 214. Also see McDowell's own arguments in support of the idea of interpreting Frege's theory as allowing ascriptions of attitudes or thoughts to objects *de re*, pp. 214–27, and 261–74.

since Evans and McDowell support the idea that such uses can be formulated within a Fregean framework, and finally, since our interpretation of Kant's uses of the same necessity follows Kaplan's interpretation, Evans' and McDowell's arguments can, and will, be cited in support of our interpretation of Kant.

In addition, as indicated just above, the logical distinction between entailment and presupposition makes a further appearance in a defense of Kant's uses of *de re* necessity. The justification of his ascriptions of necessary properties and relations to existing objects in the face of criticism leveled at it by Bertrand Russell and James Van Cleve rests on the grounds that whereas they interpret Kant as using *entailment* to state these ascriptions, Kant should rather be interpreted as using *presupposition.* The use of the latter relation alone defends Kant against their otherwise devastating criticism.

Finally, the book ends on a note of departure from Kant. In the course of the investigation of his uses of *de re* necessity it becomes clear that the interpretation of those uses contains a form of argument that is as applicable to certain *realistic* uses of the necessity as to Kant's *idealistic* uses of it. It thus becomes clear that Kant's uses can be seen as instances of that more general form of argument – a form that lends itself just as readily to a realistic interpretation of *de re* necessity as it does to an interpretation of Kant's idealistic uses of the modality. Once Kant's conception of metaphysics is seen in this light, it tends to further confirm the view in *Kant's Theory* that were it not for his belief that metaphysics can be defended only as a wholly *a priori* discipline, Kant would have little reason to embrace transcendental idealism. Therefore, for those who think that metaphysics not only must be validated in experience, as Kant himself evidently maintains with his critique of pure reason, but can also be grounded in experience, which he vigorously disputes, Kant's conception of the field will be seen as unduly narrow and his transcendental idealism unnecessary: For such a philosopher, our natural inclination to be realistic about the world doesn't have to be curbed as we pursue our intellectual interests in the field of metaphysics.

Chapter 1 – General Review

[1] Analytic Philosophy and Kant's Theory of Necessity. Analytic philosophy has leveled many challenges at Kant's theory of necessity – his account of how we can ascribe necessary properties and relations to objects, or more generally, an account of the satisfaction of a necessary condition. Some of these challenges can be answered, it is submitted here, largely in terms of techniques belonging to analytic philosophy itself. The preparation and development of that Kantian response to certain of these challenges is the primary objective of this book.

The challenges come from opposite directions, but both stem from conflict between existence and necessity. On the one hand, W. V. Quine points out that existing objects can be referred to in various ways, but only certain ways will sustain ordinarily accepted ascriptions of necessary properties and relations to the objects.[1] Other modes of reference will turn the ascriptions of the same necessary properties and relations to the same objects into falsehoods. A commitment to such ascriptions involving necessity will therefore imply a commitment to one of the preferred modes of reference. But that would be tantamount to adopting a metaphysical view of the world that sorts properties and relations according to essences (revealed through preferred modes of reference) and accidents of objects. How can such a view be supported – a question not only Quine, but Kant himself, asks of philosophers.

Kant's response to the question is that the necessity in question derives not from the nature of the objects in question, in our case, the nature of initiators – the external objects that get our knowledge started – but from the structure of the mind of the subject that intuits and conceptualizes them. The "preferred" modes of reference are simply forms of the mind that determine our intuitions of initiators. But this response only invites the challenge that comes from the opposite direction in analytic philosophy. It was originally advanced by Bertrand Russell and has recently been resurrected by James Van Cleve.[2] This is the objection that

1 W. V. Quine, "Reference and Modality," in *From A Logical Point al Point of View.*

2 James Van Cleve, *Problems from Kant* (New York: Oxford University Press, 1999), p. 37 ff. See Russell, *The Problems of Philosophy* (London: Oxford Uni-

we – human beings – might have had different forms of cognition, and in particular, different forms of intuition, from those we in fact have – something that Kant himself comes close to allowing, if he does not actually acknowledge it in just these terms. Consequently, what we find to be true of initiators according to our present modes of intuition – that is, what is true of them as appearances – might have been false. Therefore, what is true of appearances cannot be *necessarily* true. Kant's theory of necessity must therefore be mistaken.

So, both challenges demand of Kant's theory that it keep ascriptions of necessary properties and relations to appearances ordinarily accepted as true from turning into falsehoods: Kant's theory must satisfy the standard condition of *salva veritate* on possible substitutions of co-referential singular terms. Otherwise, the ascriptions cannot be necessary truths, and Kant would have no defensible theory of *de re* necessity. That is the problem that the present book will try to solve on the basis of Kant's theory of *a priori* knowledge.

The solution takes the form of a compromise. On the one hand, Kant gives the existence of initiators its due – it is independent of the mind, and in that sense it is called "real." However, the necessity of specific properties and relations that are ascribed to our appearances of initiators is due entirely to the mind, and in that sense it is called "ideal." The task for the book is to make clear how Kant resolves this conflict between real existence and ideal necessity, and it argues that he does so by reaching a compromise between them. He gives existence its due by having the *empirical* character of our intuitions depend on the existence of initiators insofar as we are affected by them. And he gives our human modes of intuition and thought their due by having the same empirical intuitions necessarily conform to those modes of intuition and the intuition conform to the modes of thought.

The compromise can largely be laid out in terms of techniques that have originated in the very analytic philosophy that has taken Kant to task for trying to resolve the conflict in the way that he did. He can thus respond to analytic philosophy in its own terms. Although the interpretation of the *Critique of Pure Reason* in Part I is meant to be a careful and rigorous preparation for the compromise of the conflict that is to come in Part II, the book is not a traditional *exegesis de texte.* Rather, although Part I is to a large extent an interpretation of both his specific

versity Press, reprinted 1957 (first edition 1912), pp. 86–87, cited by Van Cleve, ibid., p. 38.

(chapter 2 and, to a lesser extent, chapter 4) and his generic (chapter 3) notions of existence, its second part is an adaptation of techniques from analytic philosophy that allows us to understand his compromise between existence and necessity in contemporary terms.

Finally, the last portion of the book (chapters 10 and 11) amounts to a departure from Kant, and thus constitutes a secondary objective of the book. It claims that in its very defense of Kant against Russell and Van Cleve it has found a form of argument that can be applied to an oxymoronically entitled post-Kantian *realistic* interpretation of not just *de re* necessity, but of the rest of the *de re* alethic modalities as well – impossibility, possibility, and contingency. It is called "post-Kantian" because it employs a form of argument regarding the modality of necessity that is claimed to have been found in the preceding interpretation of Kant's idealistic account of *de re* necessity.

Kant's idealistic uses of *de re* necessity can therefore be viewed as not an instance of a more general form of understanding the modality, and the other alethic modalities as well, but as a competitor with at least one post-Kantian realistic interpretation of the modalities – an interpretation that is contained in chapters 10 and 11, which conclude the book's work on the modalities. As an interpretation that accords with our natural inclination to view the world realistically, it runs the risk of pushing Kant's idealistic interpretation aside as we try to make up our minds about the best interpretation of the modalities among the several that are currently available to us. If we choose a realistic one, whether the one offered here or elsewhere, for example, by Strawson or Kripke, we will have to reject Kant's very conception of metaphysics as the study of the possibility of our *a priori* knowledge of objects. As a consequence, we would not need to adopt Kant's transcendental idealism and its accompanying doctrine of faculties of the mind. Against this presumed benefit we would have to give up our defense of Kant's pursuit of his major objective – the *a priori* demonstration of the possibility of our *a priori* knowledge of objects. This in turn would entail that we reject his idea that there is a single way of studying metaphysics correctly. If that is too high a price to pay for returning to a way of understanding the world that is more natural for us, not to mention a way that is more compatible with our current pursuits of metaphysics – the realistic way, it at least makes explicit that something like a choice in involved in the adoption of a method that is suitable for studying metaphysics.

[2] The challenge from Quine. The first challenge I just outlined comes from Quine, who questions not only our ascription of necessary

properties and relations to *objects* – so-called *de re* necessity, but our ascription of necessity to entire propositions – entitled *de dicto* necessity – as well. The latter includes, but for Kant is not restricted to, the necessity Kant and most analytic philosophers attribute to propositions that they say are true by virtue of their meaning alone – *analytic* propositions.[3] Our concern in this book, however, is with Kant's theory of the first type of necessity – *de re* necessity.

Quine argues that we cannot make *quantified* statements about *any* objects – existing or abstract – which ascribe *necessary* properties or relations to the objects. These are statements in which the objects are referred to only by *variables* – expressions that refer to any object in a given domain of objects, and therefore do not of themselves distinguish any particular object from among all objects in a given domain. Consequently, the objects that are referred to by variables of quantification can be distinguished from one another only by *predicate constants* of statements. In cases in which constant *subject*, or *referential*, expressions – names, descriptions, and other expressions – for *identical* objects are substituted for the variables and where the predicates stay the same, the substitutions allow true statements to be turned into false ones, which is contrary to the condition of *salva veritate* on such substitutions (sometimes called *Leibniz's Law*, after Leibniz's principle that identical objects have all their properties in common). Henceforth, this *principle of substitution of identities salva veritate* will be abbreviated as *the principle of substitution.* Quine contends that the only things that change besides the truth values of the statements in such cases are the *manners* (or *modes*) in which the objects are referred to. Therefore, the reason that ascriptions of purportedly necessary properties and relations are ordinarily accepted as true is the manners or modes in which we refer to the objects, and not anything about the actual relation between the properties and relations and the objects themselves. Such necessity therefore has no objective ground. Or, if one assumes the contrary position, viz., that the ground of the ascriptions of the alleged necessary properties and relations is objective, since the only ground for the ascription is the manners in which the objects are referred to, the manners of reference must in some way reveal the essential nature of the objects. This position leads directly to traditional Aristotelian essentialism. So, one is either skeptical of the intelligibility of *de re* ascriptions of necessary properties and relations to objects altogether or one adopts Aristotelian essentialism. At the time of his composing

3 W. V. Quine, "Two Dogmas of Empiricism," in *From A Logical Point of View.*

these thoughts Quine thought that the choice for his readers would be easy.

Quine's famous example of a change in truth value because of a change in a mode of reference to the objects of predication is the quantified statement that there is a number that is necessarily greater than five. The numeral 'nine' and the astronomical description 'the number of planets' refer to the same object – the number nine.[4] If we substitute the numeral we get a true statement ('nine is necessarily greater than five'), but if we substitute the astronomical description we get a false one ('the number of planets is necessarily greater than five') – a violation of the principle of substitution. (How many planets there are is a contingent, not a necessary, fact, once it is decided what counts as a planet in our solar system.) Quine contends that the only difference is the difference between the numeral and the description – a difference in the mode in which the identical object – the number nine – is referred to, but obviously not anything in the object itself, since the object remains quite the same throughout the change.[5]

Quine maintains that the failure of such substitutions *salva veritate* must lead to the rejection of ascriptions of necessary properties and relations to objects, unless, that is, one is prepared to adopt traditional Aristotelian essentialism.[6] Since most of his audience is (was) not prepared to do that, they obviously must conclude, if Quine's argument is sound, that quantification into statements that ascribe necessary properties or relations to objects cannot be made intelligible. In other words, those who do not want to adopt the essentialism cannot say that there are *any* instances of necessary properties or relations.

Quine's argues that two steps lead to the unwelcome traditional Aristotelian essentialism. First, we can refer to objects of such necessary properties or relations only if the modes of our references are limited

4 Since the most recent meeting of the international astronomical association that decides such things as what constitutes a planet in our solar system changed the number from nine to eight, then, if keeping up with the latest decision of the association were to determine our philosophical uses of the idea of the number of planets, as time went on we might end up in greater confusion than we are presently in about the philosophical question of numbers having necessary properties and relations, and if so, how are they to be referred to. So, for the sake of distinctness, we'll just stick with nine as the number of planets.

5 W. V. Quine, "Reference and Modality," in *From A Logical Point of View*, p. 143 ff.

6 Ibid. p. 155.

in some way or other, and second, these limitations entail the essentialism, since a preference of some mode of reference over others is just another way of saying that certain properties or relations belonging to the preferred mode of reference are essential to the objects while all others are merely accidental. Quine takes the last statement to be just another form of traditional Aristotelian essentialism.

Kant would not only agree with Quine on the first step, but he actually takes that step when he argues that space and time are our two *a priori* modes of intuition (A38–39/B55–56, B67–68, *et passim*). Since space and time belong to objects only insofar as they are objects of our intuition, and since our intuitions of initiators must conform to our *a priori* modes of intuition, space and time can belong to initiators only insofar as they stand in relation to our modes of intuition. But so related, initiators are appearances. Consequently, space and time can belong only to appearances. Consequently, any necessity with which they can belong to initiators requires that the initiators be appearances, which in turn requires that the initiators stand in relation to our modes of intuition. To sum up, it is only insofar as they stand in that relation – it is only insofar that initiators are appearances – that initiators must be spatial/temporal.

But Quine seems to be wrong in holding that such a conclusion must lead to the tradition connected with Aristotelian essentialism. On the contrary, Kant, for example, holds that *only* the *converse* is true – that the essentialism depends on the proposition just concluded, viz., that appearances are necessarily spatial/temporal, and not conversely. In other words, merely "preferring" certain modes of intuition, though it does commit one to the ascription of certain necessary properties and relations, does not of itself commit one to Aristotelian essentialism. Spatial and temporal properties and relations are not Aristotelian essences, and therefore the objects to which they belong – appearances – are not *thereby* Aristotelian substances.

This position is confirmed by the *order* of the two propositions in the *Critique* – the Transcendental Expositions of the concepts of space and time precede the Transcendental Deduction of the categories, including, in particular, the category of substance-inherence, which is the category that involves essentialism. It is further confirmed by the *dependence* of the Transcendental Deduction on these two Expositions. In conclusion, since mere "preference" for particular modes of intuition does not of itself entail essentialism of the Aristotelian variety, and even though adherence to the principles of the substitution (of identities) and existential generalization in ascriptions of necessary properties and relations to existing objects may be based on a so-called "preference" for particular modes of intuition, the principles can still be observed in such ascriptions independently of a commitment to the essentialism.

How Kant does that is part of what is here called "Kant's compromise." Very simply, while Kant does not make the *existence* of initiators relative to our modes of intuition, that is, while he does not idealize their existence, he does make the *necessity* of the properties and relations we ascribe to their appearances depend on those same modes of intuition. It is this difference between Kant's distinct treatments of existence and necessity that constitutes his compromise of the conflict between existence and necessity.

The technique from analytic philosophy that is employed to show how Kant makes this compromise comes from an article by a philosopher of language, David Kaplan. In "Quantifying In,"[7] Kaplan proposes a novel interpretation of our use of *de re* necessity. Instead of quantifying over *objects* in existential propositions that ascribe necessary properties or relations to objects, when using constructions of *de re* necessity, we should be understood as quantifying over *expressions* for objects, where, in addition, we should understand that the expressions *necessarily* denote the objects in question. Though Kaplan limits his proposal to the relation between expressions and *abstract* objects, it is adapted in the present book to Kant's ascription of necessary properties and relations to initiators and to their appearances. A Kantian response to Quine is thus provided via Kaplan.

[3] Precedence for Interpreting Kant as In Response to Quine. There is precedence that a book on Kant takes certain of its problems from Quine. Clearly, P. F. Strawson's *The Bounds of Sense* is the classic work on Kant in analytic philosophy.[8] Yet Quine – Strawson's perennial adversary – had to have been in the back of Strawson's mind as he wrote his book on Kant. It would be hard to understand what was of concern to Strawson in the book if one were unfamiliar with his concerns with Quine. So, asking how Kant might respond to certain of Quine's prob-

7 David Kaplan, "Quantifying In," *Words and Objections: Essays on the Work of W. V. Quine*, ed. by D Davidson and J. Hintikka (Dordrecht-Holland: D. Ridel Publishing Co., 1969), reprinted in *Reference and Modality*, ed. L. Linsky (London: Oxford University Press, 1971). References to "Quantifying In" will be to this reprint.

8 P. F. Strawson, *The Bounds of Sense* (London: Methuen & Co., 1966).

lems can find its grounds in Strawson. This book is the product of concerns similar to those attributed to Strawson.

[4] The Challenge from Russell and Van Cleve. As already noted, Kant's very dependence on our modes of intuition in explaining our ascriptions to existing objects of necessary properties and relations only undoes the necessity in question, Russell and Van Cleve contend. For it is a contingent fact that we have the modes of intuition – Euclidean space and time – that we do have. Since they might have been different, what is true in the present circumstances could be false if our cognitive constitution were different.

Again a technique from analytic philosophy itself is employed to answer a challenge from analytic philosophy. The technique actually emerges from the discussion of *existence* in Part I of the book. As already indicated in the Preface, use is made of the now common distinction between the logical relations of presupposition and entailment, the latter being the so-called 'strict' or 'logical' form of the more general relation of implication, to distinguish it from the truth-functional 'material implication'.

The origin of the modern use of presupposition is located not only where it is commonly found, viz., in Frege, and following him, in Strawson,[9] but also in early Russell – well before Strawson employed it in his attack on Russell's Theory of Descriptions. The distinction between the two logical relations is then used to defend Kant against Russell and Van Cleve. It is argued that for Kant the necessity of the truth of Euclidean geometrical propositions implies that the propositions *presuppose*, but do not *entail*, that the propositions are determined by our form of outer intuition. So, if the form of our outer intuition were different, as Russell and Van Cleve suppose it might be, Euclidean geometry would not be *simply false*, but would instead be *neither true nor false.* Of course, it would not be necessarily true, either. But Russell and Van Cleve's objection to Kant is that according to his own theory of necessity, Euclidean geometry would be *simply false*, if our cognitive constitution were different from what it is. They think that the relevant logical relation in question is entailment, and this relation between Euclidean geometry and our forms of intuition *would* make Kant's theory of necessity untenable.

---

9 Strawson disavows having borrowed his early ideas, including that presupposition, from Frege, since he says he hadn't even read Frege at the time he came up with them. See *The Philosophy of P. F. Strawson*, ed. Lewis Edwin Hahn (Chicago & La Salle, IL: Open Court Publishing Co., 1998) p. 7.

The attribution to Kant of this particular use of the logical relation of presupposition is then developed as a positive account of his theory of the necessity of Euclidean geometry; yet it is merely suggested that the account can then be extended to the necessity of the principles that determine the *temporal* relations of our inner intuitions of the mind and its states and the temporal relations of events that are causally related to one another. This would entail an interpretation of Kant's uses of *de re* necessity in his theory of the necessity of arithmetic, an interpretation which is yet to be fully developed. On the other hand, the interpretation is developed as an account of Kant's theory of the possibility of synthetic *a priori* judgments that he calls *the principles of the understanding.* In particular, following the interpretation of Kant's uses of *de re* necessity in his theory of the necessity of geometry, the same interpretation is employed in an interpretation of his account of the possibility of causal judgment in general. The application of the interpretation of Kant's uses of *de re* necessity to his theory of the possibility of the other synthetic *a priori* judgments is left for a less schematic work than is intended by the present volume.

An unexpected observation can be drawn from the positive account of Kant's theory of geometry. By the mere use of presupposition and the truth-functional logical relations of conjunction and material implication, the account demonstrates how an ascription of necessary properties and relations to existing objects *must be true, if it merely has a truth value.* (This can be compared and contrasted with Kripke's formula for metaphysical necessity: if true, then necessarily true.) If the account is sound, it can be used to justify Kant's claim that the properties and relations in question would *necessarily* belong to appearances.

As indicated in the Preface, if the account is sound, it can also be used to propose a post-Kantian, *realistic* interpretation of not just *de re* necessity, but of three other *de re* alethic modalities as well – impossibility, possibility, and contingency. Accordingly, the book concludes that since we are naturally all realists in our approach to the world, we might consider returning to our natural inclinations about metaphysics and let Kant's conception of it be relegated to the study of Kant's attempt to validate distinctively *a priori* knowledge of objects and, perhaps not inconsequentially, what we might learn from that attempt in our own study of the possibility of *a priori* knowledge. Again, one of the main theses of my previous book on Kant seems confirmed: For Kant, the only knowledge the explanation of whose possibility requires transcendental idealism is our *a priori* knowledge of objects; the possibility of empirical knowledge, or experience, certainly does not require it. For Kant, if the only knowledge we were trying to account for were empirical knowledge, there would be no reason to embrace transcendental idealism: One can be quite realistic about experience.

[5] An Epistemological Criterion of Existence. As the Preface to the book explains, the criterion of objects that exist independently of us – real objects and real existences – is not the same as that offered in *Kant's Theory*. It is an epistemological and not a logical criterion of existence. The initial chapters of Part I of the book develop this new criterion of existence. Though it addresses the question of Kant's ontology from an epistemological point of view, it nevertheless picks out exactly the same set of objects that were picked out by the logical criterion of the earlier work. That is, the two criteria are co-extensive.

Finally, the epistemological criterion of existence identifies for Kant the objects of our ascriptions of necessary properties and relations in a certain way. Not that the logical criterion of existence could not have been used for the same general purpose of applying the concept of existence to objects. But the advantage of the epistemological criterion would have been lost. For now both existence and necessity can be understood in epistemological terms. As will be argued in chapter 3, since Kant's generic notion of *existence* depends on the general *epistemological* notion of an intuition – whether sensible or intellectual, and since space, time, and the categories are epistemological concepts for him, he can use both sets of terms *epistemologically* in his ascriptions of *necessary* spatial, temporal, and categorial properties and relations to appearances. That is something that could not be readily done in the previous book.

Kant's epistemological criterion of existence in general has a further advantage over the logical criterion of existence. The ontology of his *theory* of sensible intuition is simply a species of existence in general. Not only are these the objects that get our *knowledge* started, i. e. initiators, but the objects with which Kant begins his *theory* of our knowledge – in both editions of his Introduction to the *Critique*. To repeat, all our knowledge begins, he says there, with experience, and experience entails for Kant that we are affected by objects that are given to us – in space and time. So, the relation of the specific, i. e. sensible intuition, to the general, i. e. intuition in general, whether sensible or intellectual, is the logic of the relation of chapter 2 to chapter 3 of the book. So, from its early chapters forward, the book interprets the *Critique* from an epistemological, instead of a logical, point of view.

However, as the Preface to the book has already indicated, and as the previous section of this chapter already suggests, following the logical bent of the earlier work, this book also offers a *logical* distinction between initiators and their appearances. Only this time it is not the transcendental logic of the *Critique* itself, but ordinary formal logic, the one Kant called *general logic.* Yet it is not quite the general logic familiar to Kant and his contemporaries. It is rather the standard logics of truth-functional sentential connectives and the predicate calculus plus the logical relations of entailment and presupposition that are prominent in philosophy of language, the distinction between which will be explained in even greater detail in chapter 5. Again, just as the earlier transcendentally logical criterion of Kant's ontology is co-extensive with the new epistemological criterion of the ontology, the new logical distinction between appearances and initiators picks out objects that can also be distinguished according to the new, epistemological criterion of Kant's ontology.

The new epistemological criterion will be elaborated in chapter 2 and developed in chapters 3 and 4, while the new logical distinction between appearances and initiators will be introduced in chapter 5 and developed in chapters 8 and 9. Chapters 10 and 11 then employ a logical interpretation of certain of the alethic modalities that it is claimed has been found in the interpretation of Kant's uses of the modalities that have been given in the previous chapters of the book. This frees us to pursue our investigation of our uses of these modalities independently of whatever interest we may have in Kant's uses of them.

[6] Summary. Starting with Kant's opening thought in the Introduction to the *Critique* – how our knowledge begins, the book (a) provides a new dual-aspect *epistemological* interpretation of Kant's transcendental idealism. The single set of objects are identified epistemologically as the external objects that get our knowledge started – initiators. These are then understood as objects that in *affecting us* (in our *a priori* modes of intuition) are *given to us* (as determined through these modes). It is then argued that Kant separates the two aspects of initiators – affecting us and being given to us – and employs them as distinct grounds for positing two sets of distinctively Kantian theoretical objects – *appearances* and *things in themselves.* He does that, however, only to account for the possibility of our *a priori* knowledge of objects, i. e. *appearances*, and not to advance either a theory of *empirical* knowledge (or experience) or a theory about reality – what he would call an 'ontology'.

The book then (b) proposes a criterion for Kant's *generic notion of existence*, of which the existence of initiators in (a) is a specific instance. It does so by finding a notion that covers the existence of objects of *intellectual* as well as *sensible* intuition – something that seems to have so far eluded Kant commentators.

The analysis continues with (c) an interpretation of the relation Kant envisaged between human knowledge and the existence of the objects that we can know. As already stated, the relation is an adaptation of the logical relation of *presupposition* discussed above, namely, the relation that was introduced into analytic philosophy by Frege, then by Russell, and more recently, following in Frege's path, by Strawson.[10] But Kant's theory of intuitive reference does not fit neatly into the semantics of any of the three of them. Consequently, neither Russell's attack on Frege's theory of sense and reference via the Theory of Descriptions nor Strawson's retaliatory attack on Russell's Theory of Descriptions keeps Kant from sharing certain features belonging to one side or the other of the controversy.

The oscillation between the two semantics allows (d) a *quantificational* interpretation of certain of Kant's uses of *de re* necessity with regard to the ascription of necessary properties and relations to objects – an interpretation whose formula is borrowed from David Kaplan. Being quantificational, it entails that Kant's uses be understood in a way that would be compatible with a partial analysis along the lines of Russell's Theory of Descriptions. The analysis would be only *partial* because the adaptation of Kaplan's interpretation to Kant contains Frege's and Strawson's idea that our *ordinary* references to objects presuppose rather than entail the existence of the objects. (Russell employs presupposition only in regard to the references to objects of logically proper names, not definite descriptions, of which he considers our ordinary proper names of objects to be mere abbreviations.) Besides allowing quantification into necessity contexts, Kaplan's formula allows Kant's ascriptions of necessary properties and relations to remain in logical relation to the existence of objects. Therefore, though existence is real for Kant and necessity, ideal, the two can be related through the quantificational technique provided by Kaplan.

Although presupposition may not appear to be essential to this Kaplanesque *interpretation* of Kant's uses of *de re* necessity, it is obviously essential to (e) the *justification* of those same uses, and therefore must be essential to the interpretation as well. Kant's theory of *de re* necessity is explained on the grounds that his ascription of the necessary properties and relations of space, time, and the categories to appearances does not

10 But see the previous footnote.

entail that we intuit the objects that appear – initiators, but instead only *presupposes* that we do. Therefore, the existence of the initiators is presupposed, but not entailed, by the ascription in question.

These five theses – (a)–(e) – are the major claims that the book puts forward in its interpretation and defense of Kant and are therefore the conclusions of the arguments that constitute the book's Kantian contents.

In addition, as already noted at the conclusion of the opening section of this chapter, the book also has a secondary objective, which is independent of its interest in Kant. It proposes a realistic interpretation of the *de re* modalities that it is claimed has been found within the preceding interpretation of Kant's own uses of the modalities. Kant's uses thus become instances of the more general understanding of the modalities and might thus be viewed more as an object of historical interest than as a lesson in the supposedly only way to conceive metaphysics, unless, that is, one sees Kant's idealism as a guide toward validating claims of our having *a priori* knowledge of objects and insists on interpreting our uses of the modalities as claims belonging to such knowledge.

Part I: Existence

Chapter 2 – How Our Knowledge Begins

[1] Introduction. An interpretation of Kant's theory of knowledge faces, *inter alia*, two questions: Which objects that affect our senses get our knowledge started? and How can the answer to that question avoid certain profound problems with the answer given by the *dual-object* interpretation of transcendental idealism, viz., that the objects in question are things in themselves? Without an answer to the first question, we cannot understand Kant's explanation of how our knowledge begins, and without an answer to the second, we cannot be satisfied with the answer to the first. This chapter first frames both questions and then examines two sets of answers to them – one set dealing with the currently popular dual-aspect interpretation of transcendental idealism, the other with the dual-object interpretation of the idealism. It concludes that we can understand how our knowledge gets started by objects only if the objects are understood according to a *dual-aspect* interpretation of the idealism, but not the one that is currently popular.[1]

[2] The Identical Object that is Given to Us and that Affects Us. Kant opens the Introduction to the second (B-) edition of the *Critique* with the famous statement that all our knowledge begins *"in the order of time"* with experience (B1).[2] He explains that we give the name *experience* to our knowledge of objects that comes from our "working up" (B1)[3] the sensi-

1 Most prominently represented by Gerold Prauss, *Kant und das Problem der Dinge an sich* (Bonn: Bouvier Verlag Herbert Grundmann, 1974, chap. 2, and by Henry E. Allison, *Kant's Transcendental Idealism: An Interpretation and Defense* (New Haven: Yale University Press, 1983), esp. 247 ff. Allison has revised his position in a second edition of the book, 2004, but has reiterated his methodological interpretation of Kant's idealism as recently as 2006, in his article, "Transcendental Realism, Empirical Realism, and Transcendental Idealism," in *Kantian Review* (Cardiff: University of Wales Press, 2006) Volume 11, pp. 1–28. The controversy over the correct interpretation of Kant's idealism continues elsewhere, as well. See, for example, Lorne Falkenstein's "Critique of *Kantian Humility*," in *Kantian Review*, Volume 5. pp. 49–64, and Rae Langton's reply in defense of her book in the same volume, pp. 64–72.

2 "***Der* Zeitnach**" In the first paragraph of the Introduction Kant identifies the knowledge in question both as "our knowledge" and as "knowledge in us."

3 "*verarbeiten*".

ble impressions that, he later says, we get from the objects' affecting our senses (A19/B33).

Before going further, it should be made explicit that what Kant means by *experience* is not limited to the individual perceptions that it comprises, but to "connected (*verknüpfte*) perceptions" as well (B161). The connection is made in accordance with the categories (ibid.). Kant thus means by *experience* whatever possible perceptions and their objects that extend out from whatever actual perceptions in which experience might originate as well as the original, actual perceptions. Again, the extension is determined by the categories. Since natural laws for Kant are to be explained as the rule-governedness that the categories prescribe to appearances (B159, B164), the reach of experience extends as far as the empirical laws of nature will allow. This way of understanding experience provides considerable ground for his claim that "[o]ur knowledge of the existence of things reaches, then, only so far as perception and its advance according to empirical laws can extend" (A226/B273).

Returning now to the actual passage at A19/B33 cited just above, we should note that it states, not merely that objects must affect our senses, if we are to get sensible impressions, but that the objects that do so are the very same objects that we experience. Kant's language is unmistakable in this regard: "[intuition] takes place only insofar as the object is given to us; but this in turn, is possible only if *it* affects the mind in a certain way" (A19/B33, Guyer-Wood translation, my emphasis). The anaphoric use of the relative pronoun 'it' in this passage entails that the object that affects the mind is the identical object that is given to us through the intuition.[4]

The neglect of this identity – between the object that is *given to us* through the intuition and the object that *affects the mind in a certain way*, i.e. *through the same intuition* – has unfortunately led to an unnecessary and distracting controversy in the interpretation of transcendental

4 Before proceeding further, it should be mentioned at this point that this independence between the object that affects us in a certain way and the way in which we are affected by it will be spelled out further in sections 5 and forward below. This will be done in terms of the notion of an 'external object' – external to our sensible cognitive constitution – a notion that I will explain in terms of details from Kant's own system in those later sections of this chapter. It is the notion I there use for the identical object that we have been talking about thus far. There is subsequently a fuller discussion of the importance of this identity between the affecting object and the given object for my interpretation of the external object especially in sections 8 and 9 below.

idealism. It is the dispute between double-aspect and double-object interpretations of the idealism, as sections 8 and 9 below will attest.

If we now return to the original passage from the Introduction to the *Critique,* we find that although Kant says that all our knowledge begins in time with experience, he also says that "not all of it arises out of experience" – some of it comes from "the faculty of knowledge itself" (B1), including, he later says, the aforementioned ways in which we are affected by objects – space and time, since that is how we experience the objects as being (A39/B56 ff.). The knowledge that accordingly comes from only the "faculty of knowledge itself" is *prior* to experience and is thus said to be *a priori.* Since, as part of our total knowledge, our *a priori* knowledge begins in time *with* experience, Kant must mean by *a priori* here that it is *logically,* not *temporally,* prior to experience; that is, *a priori* knowledge is prior to experience in what we might call *the order of explanation,* but not *the order of time.*

So, when he says in the Introduction that not all of our knowledge "arises out of experience," he is talking about our *a priori* knowledge and he must be saying that it arises out of a condition that is *atemporally antecedent* to experience, not a condition that is *temporally* antecedent to it. It is in this atemporal sense that the condition is said to precede experience. It does so for Kant as a condition of what he calls the *possibility* of experience.

[3] Space and Time Either as Knowledge or as Belonging to Knowledge. It doesn't need repeating that Kant speaks of space and time variously as modes and as forms of sensibility and of intuition, which serve as "principles of *a priori* knowledge" (A22/B36), as sources or grounds of knowledge (A50/B74), as forms of appearances, and, finally, as objects (especially at B160n). But what I want to explore here is the less common topic of Kant scholarship, either (A) that space and time for him are *of themselves* knowledge or the related idea (B) that space and time belong to knowledge, though they are not of themselves knowledge. Since we have already established in the preceding section that in the order of time *all* our knowledge begins with experience, it would follow that either as *of themselves* knowledge (option (A)) or as *belonging to* knowledge, i.e. as part, or elements, of the knowledge to which they belong (option (B)), space and time begin *in time* and that either as (A) or as (B) they, like the rest of knowledge, do so only *with* experience.

Beginning with option (A), three passages from the *Critique* support it. The first is from the Aesthetic. He says that our *a priori* knowledge of space and time is "called pure intuition" (A42/B60). By conversion by limitation, it follows that pure intuition is knowledge, and since space and time are pure intuitions, it would follow that they would be knowledge. According to this passage, knowledge and object would be identical.

The second passage is also from the Aesthetic, but only from the second edition of the *Critique.* At B44 he contrasts sensations with intuitions, and says of the former that they "do not in themselves allow any object to be known, least of all any *a priori* knowledge" (Guyer-Wood translation, my substitution of "known" for their "cognized"). The contrast with sensations suggests that Kant is implying that intuitions "in themselves" do allow an object to be known. But the passage is strong enough to warrant only a suggestion in support of the idea that intuitions by themselves are knowledge, since it could be read that Kant intends to withhold knowledge from intuitions "in themselves" as well.

The third passage is from the Logic, in particular, the Dialectic. At the *Stufenleiter* – the classification of representation in general (*Vorstellung überhaupt*) (A320/B377) – Kant mentions intuition as one of the two types of "objective perception" (*objective Perzeption*) that he says is knowledge (*Erkenntnis*), the other being concept. So, in this important passage he actually calls intuition 'knowledge'. Since the expositions (*Erörterungen*) of the concepts of space and of time assert that space and time are intuitions, they would *of themselves* be knowledge, according to the *Stufenleiter.* The *Stufenleiter* thus gives grounds for adopting option (A), and object and knowledge would again be identical.

It might be felt that these three passages must be slip-ups on Kant's part, since in the Analytic he is explicit that knowledge requires concepts as well as intuitions, i. e. that only option (B) is acceptable. He speaks of both intuitions and concepts at the beginning of the Analytic as "elements of *all* our knowledge" (A51/B75, my emphasis). Similarly, at B137, he states that the manifold of an *a priori* intuition must be united through the unity of apperception, if it is to constitute knowledge. So, if the understanding brings an *a priori* synthetic unity to the manifold of space or of time, that would seem to be enough for either of them to belong to knowledge, if either one is determined by a suitable concept, and thus determined by the synthetic unity of apperception.

Option (B), however, might seem to suggest that a distinction needs to be drawn between the *objects*, space and time, and our *knowledge* of them, whereas option (A) precludes the distinction, that is, it treats the objects and our knowledge of them as identical, since, as already noted, space and time are intuitions, and if intuitions are of themselves knowledge, so are space and time. The famous footnote to B160 provides grounds for the distinction between object and knowledge. For it speaks of space and time as "formal intuitions" whose respective unities, which are discussed in the Aesthetic, "as a matter of fact . . . presuppose" *only* a *synthesis* of the understanding, but not a *concept* of the understanding. Consequently, we could let the *objects* space and time be constituted by the understanding's *mere* synthesis of the manifold of the *intuitions*, or *objects*, space and time, and let the *knowledge* of these intuitions, or objects, be constituted by the understanding's synthesis of the manifold *according to its concepts*. The suggestion, based on the distinction just drawn, seems to entail an opposition between options (A) and (B), and thus two opposing answers to the question of whether space and time are knowledge, since (A) precludes the distinction between object and knowledge and (B) appears to suggest it.[5]

A passage from the Preface to the second edition of the *Critique*, however, though it falls under option (B), can help resolve any apparent opposition between the two options.

> Since I cannot rest in these intuitions if they are to become knowledge[s] (*Erkenntnisse*), but must relate them as representations to something as their object, and determine this latter through them, either I must assume that the *concepts*, through which I obtain this representation, conform to the object, or else I assume that the objects, or what is the same thing, that the *experience* in which alone, as given objects, they can be known, conform to the concepts (Bxvii, amended translation).

The passage makes it clear that a concept's determination of an intuition and that of its object are one and the same thing. As the intuition becomes knowledge, the object becomes known: The concept's determination of one just *is* its determination of the other. It is natural, therefore, for Kant to speak of the object and the experience of it as "the same thing."

Though the passage is clearly speaking of objects of experience, its reasoning is just as applicable to the relation between objects of our *a priori* intuition, space and time, and our *a priori* knowledge of them. The relation is clearly identity. The footnote at B160 would then be interpreted as follows. Yes, in the Aesthetic, the understanding only provides the *synthesis* of the manifolds of space and of time, whereas in the Logic it provides their synthetic *unity*. So, the intuitions, or objects, space and

---

5 I engage in an extended discussion of the note at B160, one related to the present discussion, in *Kant's Theory*, pp. 219 ff.

time, in the Aesthetic are not yet knowledge. This counts against option (A). But option (B) doesn't actually support the distinction between objects, or intuitions, and knowledge that it seems to suggest. Since a concept's determination of the manifold of an intuition that makes the intuition knowledge is at the same time the concept's determination of the object that is now known, the object and the knowledge are identical. But this is a result of option (A) as well. So, the difference between the two options, though significant, doesn't affect the question whether space and time are knowledge. Whether the intuitions, space and time, *of themselves* are knowledge or are knowledge if and only if they are determined by concepts – one way or the other – they are nonetheless *knowledge.*

The upshot of our resolution of the issue between options (A) and (B) is that since Kant has already stated that all our knowledge begins in time with experience, and since we have now established that space and time are knowledge, whether of themselves or as determined through concepts, *they begin in time,* and given the conditions laid down in the previous two sections, they do so only *with experience.* Consequently, when Kant states that space and time are *a priori*, he must mean that they are *logically*, not *temporally*, prior to experience. And again, to reiterate, since they are *knowledge* and *all* our knowledge begins *in the order of time* only with experience, they also occur *in time*, but *with* experience, not prior to it.[6]

[4] Space and Time as Ways in which We Represent Objects that Affect Us. Not only do space and/or time occur *in time* as being of themselves or as belonging to our *knowledge.* They also occur in time as qualities belonging to the particular *objects* that exist in space and/or in time, i. e. appearances. That is, since all the appearances to which they belong occur in time, so do the space and/or time that belong to them.

But the term 'quality' is used here only in a sense based on Kant's distinction between intuition (and therefore space and time) and concept (and therefore the categories) that is drawn at B136a. They are *not* qualities in the sense in which they can be possessed *in common* by different objects – a sense in which the objects can be said to be identical in respect of the quality, as, for example, human beings are identical in respect of their being human. Again, the sense in which I am using 'quality' is

6 I would like to thank Robert Howell for useful comments on an earlier draft of the preceding ideas about the distinction between space and time as objects and as knowledge.

not the sense in which the *unity* of the class of objects that have the quality would be what Kant would call an "analytic unity" (B138). Finally, with respect qualities in the sense in which I am again *not* using the term, all the objects having such a quality can be said to be *instances* or *exemplifications* of the quality. Rather, I am using the term in the sense in which a class of objects are related to each other as parts of a whole (B136a). In this sense, it can be said that the objects exist in the quality in question. In the case of space and time, however, Kant maintains that they are independent of the objects that exist in them. Similarly, any particular space or time in which an object exists would be independent of the object. Nonetheless, the object would play an essential role in the identification of the particular space or the particular time in which the object exists. In that manner, the space or the time in which the object exists would *not* be independent of the object. Thus, every external object exists in a part of or a particular space and every event exists in a part of or at a particular time. Where, in Kant's terminology, objects can only be *thought* in terms of qualities that can be *instantiated* or *exemplified*, they can instead be *given* in terms of the particular spaces and times in which they exist, where the particular spaces and times involved enjoy both the independence and the dependence just described.

If we think of space and time as qualities in the second, but not the first, sense just given, the spaces and times in which objects exist can be viewed as what have been considered "particularized qualities."[7] Although the more common conception of so-called particularized qualities has been applied to qualities in our *first* sense of the term, in interpreting Kant, I am applying it to qualities the second sense.[8] Consequently, space and time occur in time both as *knowledge*, whether according to option (A) or (B), and as *particularized qualities* of *objects* that exist in space and/or in time, where, again, "particularized qualities" is used in the sense just explained. The point of this analysis is that we again find an identity between knowledge and object as known.

---

7 Following Strawson's brief discussion of them in *Individuals*, space and time as belonging to objects existing in space and time (i.e. as belonging to particulars) might be called "particularized qualities." *Individuals* (London: Methuen & Co., 1959), pp.168–9 n.

8 I discuss Kant's uses of 'qualities' in a complementary way in *Kant's Theory*, pp. 78–82. This Kantian use of "particularized qualities" is largely in agreement with Houston Smit's Kantian "intuitive marks" in "Kant on Marks and the Immediacy of Intuition", *The Philosophical Review* Volume 109, No. 2 (April 2000).

The identity between knowledge and object as known gives a somewhat circuitous route to a certain proposition that is more commonly recognized as a straight-forward conceptual truth in Kant's system. It is that, with respect to objects that affect us, we must represent them in the *a priori* ways in which we are affected by them (B 41, A26/B42, A44/B61, B67 ff.). Consequently, we must represent such objects as spatial and/or temporal – that is how they must exist in our representations, or how we must represent them "as being" (A42/B59, Kemp-Smith and Pluhar translations, or, alternatively, according to the Guyer-Wood translation, represent them "to be"). Since Kant maintains that they must be given in experience if they are to be known (Bxvii *et passim*), and since he also says that they can be given in experience only through representations that are intuitions, it follows that for Kant objects that affect us must exist in space and/or in time, if they are to be known.

Whereas space and time are the ways in which we can be affected by objects, and thus the ways in which we can represent the objects as being, and whereas both space and time are also knowledge, whether they are considered according to option (A) (as of themselves knowledge) or (B) (as elements of knowledge as something that is combined), only *time* is *both* the single way in which all our knowledge begins in time *and* one of the two ways in which we must represent the objects that affect us as being, and thus *as being in time.* In other words, time alone provides a single framework of location for both our *knowledge* of objects and the *objects* of our knowledge.

[5] External Objects Get Our Knowledge Started. Kant not only explicitly states that experience depends on objects' affecting our senses in order to get started, but if we add a certain assumption, he also implies the same thing when he asserts that all our knowledge begins in time with experience. The assumption is that *experience cannot start itself.* [9] If we add the assumption to Kant's statement that experience depends on objects' affecting our senses in order to get started, it follows that something external, and hence logically prior, to experience gets it started. This ex-

9 Manley Thompson, "Things in Themselves," in *Proceedings and Addresses of The American Philosophical Association* (Newark, DE: The American Philosophical Association, 1983) pp. 36, 40, when discussing C. S. Peirce's "secondness" in relation to Kant's things in themselves, speaks of their mutual concern with "our sense of being in a world we never made." As will be made clear in section 9 below, I think both Peirce and Thompson are mistaken in their view that Kant's thing in itself is his primary term for expressing "our sense of being in a world we never made."

ternal something is none other than *objects' affecting our senses. In the sense in which the objects that are involved in our senses being affected by objects get experience started, they, too, are external to experience.* Consequently, they cannot be understood in terms of our *a priori* principles of knowledge (including space and time) and the *a priori* knowledge that "flows"[10] from the principles; otherwise, our knowledge *could* get itself started. Following a decision announced at the very beginning of the Preface to this book, which was made to avoid repeating a lengthy expression throughout the book, we will (continue to) call external objects that gets our knowledge started *initiators.*

Before proceeding, it should be made clear that the fact of our knowledge getting started is a contingent fact, and hence a fact that might *not* have happened as well as one that did happen. In case our knowledge *doesn't* get started, we would not have any experience, and consequently, space and time would be empty – nothing would exist in either one, since objects exist in space and time only if they are given to us in experience. Similarly, it is a contingent fact about initiators that they get our knowledge started, and hence it is a fact about them that might *not* have happened as well as one that did happen. So, although these external objects are *understood* as initiators, it nonetheless remains a contingent fact about them that they get our knowledge started. (There is nothing wrong with using a contingent fact about something as a way of identifying it.) Consequently, that contingent fact about them is part of our understanding when we understand them as initiators. Therefore, we understand them as being distinct from the knowledge that they start, which is precisely what we mean when we understand them as objects that are external to experience, and thus as *external objects.*

[6] Transcendental Idealism. As I have already noted in the Preface and in chapter 1, not to mention my earlier book on Kant, he also famously states that the problem he is attempting to solve in the *Critique* is that of the possibility of *a priori* knowledge. As part of his theory of that possibility, which Kant calls *transcendental idealism,* he entitles the objects of *empirical* knowledge (i. e. the objects of experience) *appearances.* They include material objects of ordinary experience and the objects of

10 "*herfliesen*" B 40. The word is used here in regard only to the way in which geometry is said to "flow" from space; Kemp Smith trans. Kant also uses the same word, however, at A136/B175 in relation to the way that synthetic *a priori* judgments are said to "flow" from space, time, and the categories.

(physical) science,[11] and constitute one half of the traditional dichotomy of transcendental idealism.[12] The other half consists of *things in themselves.* These are objects that are understood through their properties and relations to other objects that are logically prior to both our *a priori* and our empirically conditioned senses. Consequently, they are external to experience.

[7] Dual-Aspect vs. Dual-Object Interpretations of Transcendental Idealism. As already noted in sections 1 and 2 above, two interpretations of the *traditional dichotomy* of transcendental idealism have recently become prevalent.

The prevalent *dual-aspect* interpretation holds that transcendental idealism is really about just methodology and not ontology: Things in themselves are merely objects of thought *that makes no ontological commitment.* Allison has just recently reiterated this methodological approach as an anti-metaphysical interpretation of the idealism.[13] The prevalent *dual-object* interpretation, on the other hand, belittles its adversary for distorting Kant's idealism to the point that it becomes "anodyne," and perhaps the dual-aspect interpretation does so out of a concern to make Kant's idealism more appealing to the band of empiricists that now prevails in the academy, the dual-object commentators included among them.

[8] The Prevalent Dual-Aspect Interpretation of Initiators as *Appearances.* Since the thought of things in themselves on the prevalent, i.e. methodological, dual-aspect interpretation does not imply the existence of such objects, it is difficult to understand how they are supposed to get knowledge started by affecting our senses.

For example, Allison interprets Kant as speaking this way about the object that affects our senses "in a transcendental context."[14] But he seems to have mistaken Kant's view that the thought of the thing in itself lacks a corresponding *intuition*, and hence a *use* (*Gebrauch*) or an *application* (*Anwendung*) to an object, for the view that the thought does not imply that the object exists at all (being a mere *methodological* no-

11 Kant's treatment of the knowing subject is largely, but not entirely, omitted from my discussion (see reference to B67 ff. in section 4 above), even though my analysis could be applied to it, *mutatis mutandis.*

12 See section 10 below for further reference to the "traditional dichotomy" of transcendental idealism.

13 See his, "Transcendental Realism, Empirical Realism, and Transcendental Idealism," p. 2 ff.

14 Henry E. Allison, *Kant's Transcendental Idealism*, first edition, p. 250, *et passim.*

tion). Clearly, Kant says that since a thing in itself is thought as an object, and since such thought is always possible through the categories, and since existence is a category, the thing in itself can be thought to exist. When he goes on to say that it is so thought only "problematically," and not "assertorically," he means, I would submit, that there is no corresponding intuition for its use or application to an object, not that it lacks a commitment to the existence of the object so thought (B309 ff.).

Indeed, in the next chapter of this book that very stance is attributed to Kant. It claims that Kant's criterion for the use or application of the concept of existence is that the object thought to exist (through the concept, or category of existence) satisfies the condition that it be an object of an intuition. In the present case, reference to the criterion of existence would give the result that things in themselves while *thought* to exist – being thought through the category of existence as well as other concepts – are so thought only *problematically*, because there is no intuition of the object corresponding to the concept, and thus to the thought. It is an object of thought alone, but thought to exist nonetheless.

It would seem, therefore, that for the prevalent dual-aspect interpretation of the object that gets our knowledge started, i. e. the initiator, that object should be the identical object mentioned at A19/B33 that was introduced into our discussion in section 2 above, viz., the *appearance.* But there are *three* separate reasons that appearances cannot be the initiators, or the identical objects in question, and thus three reasons that appearances are not up to the task of getting our knowledge started. I say this even though the prevalent dual-aspect interpretation of Kant's idealism has in its favor what I consider the correct, *general* view, that it is an *identical* object that *in affecting us is given to us.* I am simply saying that it is wrong to interpret that object as the *appearance.* (We should note in passing, and as was also be mentioned in section 6 above, that the reasons I am giving against understanding the identical object as the appearance also keep the physical, or material, object from being taken as the identical object, since for Kant physical objects are appearances (e.g. A45/B62 ff.)).

The First Reason. The first reason for rejecting appearances as the initiators consists of the following argument. First, since the objects must be external to experience, and since *ex hypothesi* they are appearances, which, in this instance, would consist in the hypothesis that they are material objects, it follows that material objects are external to experience. But material objects exist in space and time. We have already noted in the conclusion of section 4 above, however, that not only do certain *ob-*

jects exist in space and time, but as belonging to, or part of, the objects that exist in space and time, the space and time in which the objects exist must also occur in time – as so-called "particularized qualities," in the sense explained in section 4. So, on the dual-aspect interpretation of the *initiator*, space and time, as the ways in which the initiators exist, can occur before the occurrence of experience. This is obviously the common sense view of material objects and of space and time.[15]

We have also seen in section 3, however, that space and time are *knowledge* for Kant (which is *not* the common sense view of space and time), and section 2 states that for him knowledge begins in time *with* experience. So, *as knowledge*, space and time, whether as intuitions *of themselves* (option (A)) or as intuitions that are determined through concepts (option (B)), occur in time *with* experience, not before it.

These propositions present the dual-aspect interpretation of the *initiator* with a dilemma. Space and time either are or are not knowledge. If they *are* knowledge, initiators cannot exist in them, since only the objects (and thus also the ways in which they exist), but not knowledge, are external to experience, in the sense explained in section 5, namely, only the objects, but not knowledge, are both logically prior to experience and can get it started. In that case, initiators can't be appearances, since appearances exist only in space and time. But initiators can't be things in themselves, either, since the latter are objects that are *not* thought to exist – at least on the dual-aspect interpretation that is currently prevalent. So, the dual-aspect interpretation of transcendental idealism currently prevalent comes up empty-handed with an interpretation of the *initiator.*

On the other horn of the dilemma, if space and time are *not* knowledge, initiators *can* be appearances. But this interpretation of the objects would have to conclude that Kant has no defensible theory of the possibility of *a priori* knowledge, since his taking space and time to *be* knowledge, namely, *a priori intuitions*, according to either option (A) or option (B), is essential to that theory. In a word, Kant would have to be so interpreted that he did not have a defensible theory that space and time are *a priori intuitions*, which provide principles of *a priori* knowledge.

(This horn of the dilemma is grasped by P. F. Strawson as the only way of interpreting the external object and transcendental idealism,

15 This interpretation would be the so-called "neglected alternative." See Paul Guyer, *Kant and the Claims of Knowledge* (New York: Cambridge University Press, 1987) p. 363, *et passim*, for a discussion and embrace of the alternative.

even though it disintegrates into unintelligibility. Again, he contends that not only is Kant's concept of the thing in itself unintelligible, but it is unnecessary in an interpretation such as his own that makes the work both intelligible and largely defensible. He takes the admirable portions of the *Critique* as constituting an essay on the fundamental structure of empirical knowledge, or experience, an essay he reconstructs in his book. He argues, however, that a *faithful* (but incoherent) interpretation of Kant entails a dual-*object* view of the *Critique.*)[16]

The dilemma for the currently popular dual-aspect interpretation of transcendental idealism – that the idealism is only a methodological claim about objects – is then the following: Either it has no interpretation of the *initiator* or Kant has no defensible theory of the possibility of *a priori* knowledge.

Before proceeding, it will be clear from each of the remaining two reasons that they apply to *any* attempt to interpret the identical object as the *appearance*, and not just to the currently popular version of that interpretation. That is, none of the following reasons hold *only* for the prevalent dual-aspect interpretation. The reasons appear at this point in our discussion of the prevalent dual-aspect interpretation because this particular interpretation of the initiator suffers no less from the force of these reasons than does any other interpretation that takes the initiator to be the *appearance.*

The Second Reason. The *second* reason that the dual-aspect interpretation of the initiator as the *appearance* is unacceptable is that it commits a logical fallacy. In the Transcendental Aesthetic, an object's affecting "the mind in a certain way" is introduced as a necessary condition of the possibility of the object's being given to us. As already noted in section 2, it is explicit in the text that it is one and the same object – the object that is given to us is the object that affects the mind. This particular consideration, of course, again counts *in favor* of the dual-aspect interpretation of the external object, since on that interpretation the given object and the one that affects the mind are indeed one and the same object – the appearance. Let us next allow that Kant's reference to an object's affecting the senses in the Introduction of the *Critique* is a case of an object's affecting the mind, or affecting "us," as that notion is introduced in the Transcendental Aesthetic. We can do so on the grounds (1) that, as already noted, the object's affecting the mind is a necessary condition of an object's (actually, the *same* object's) being given to us, and (2) that

16 P. F. Strawson, *The Bounds of Sense* (London: Methuen & Co., 1966).

Kant also holds that there is a connection between an object's being given to us *simpliciter* (the Transcendental Aesthetic) and its being given to us in experience (the Introduction) (Bxvii), and hence its affecting the senses (the Introduction).

Now, the second difficulty for the dual-aspect interpretation that takes the appearance as the initiator is that appearances are *understood* as objects that are given to us and therefore cannot be employed in an *account of the possibility* that an object is given to us. Otherwise, Kant would commit a *petitio.*[17] That is, a condition of the possibility of an object's being given to us must be understood independently of an object's being given to us, if the condition is to explain that possibility. However, since an appearance is *understood* as possibly being given to us, and since it is part of the understanding of the initiator that it affects the mind, if the initiator is *understood* as an appearance, the possibility of an object's being given to us would be *understood* to involve its affecting the mind. In that case, therefore, its affecting the mind could not account for the possibility of its being given to us.

The Third Reason. The *third* reason that appearances cannot be the initiator is that it is *incoherent* to suppose that appearances can instantiate the concept of the identical object that is mentioned at A19/B33, i. e. the initiator.[18] The incoherence I am alleging can be demonstrated through an argument that employs Frege's theory about the informativeness of identity propositions.

1. At A19/B33 Kant asserts the identity proposition (henceforth called *the identity proposition*) that the object that affects the mind is the object that is given to us.

---

17 This is a criticism dating all the way back to F. H. Jacobi, a younger contemporary of Kant's. For an early Anglophone discussion of the fallacy in taking the external object to be the appearance and its relation to Kant's theory of affection, see Norman Kemp Smith, *A Commentary to Kant's 'Critique of Pure Reason,'* (New York: The Humanities Press, 1918, reprinted 1950) p. 613 ff, who discusses the issue in terms of the "phenomenal" object.

18 In his latest version of his interpretation of transcendental idealism, Allison continues to take the identical object of his two-aspect interpretation of transcendental idealism as the "empirical" object. See his, "Transcendental Realism, Empirical Realism, and Transcendental Idealism," in *Kantian Review* (Cardiff: University of Wales Press, 2006), p. 1. However, at least at one place in his original version of his interpretation, Allison seems to recognizes the necessary independence between us and the object that affects us in a certain way when he takes the object to be the thing in itself. See his, *Kant's Transcendental Idealism*, first edition, pp. 247 ff.

2. The identity proposition is *not* of the logical form A = A nor is it of this form if it is supplemented with suitable definitions. In terms of the analytic-synthetic distinction, the identity proposition is not analytic. Kant does not understand by "affects the mind" "given to us", nor, conversely, does he understand by "given to us" "affects the mind." Accordingly, things in themselves can affect the mind without being given to us and mathematical objects can be given to us without affecting the mind.
3. That an appearance is given to us is analytic (from the initial explication of an appearance as the undetermined object of an empirical intuition at A20/B33–4, and its implications).
4. If in the identity proposition the object that affects the mind is the appearance, the identity proposition is analytic (from 3).
5. In the identity proposition, the object that affects the mind is the appearance (assumption for reductio).
6. The identity proposition is analytic (from 4 and 5).
7. 6 is false (from 2).
8. 5 is false (from 4 and 7).

The First Objection to the Argument. It might be objected that the deck has been stacked against the appearance as the affecting object, and hence as the initiator. For the identical object in question is not just the object that is given to us, but the object that is given to us *through an intuition.* This caveat would eliminate mathematical objects from the justification of proposition 2, since mathematical objects are given to us through a *synthesis* of the understanding, and not through an intuition (cf. B160a), or at least it can be so argued. [19] It is for that very reason that in Kant's system mathematical objects cannot be said to exist. Consequently, they should not count toward considering the identity proposition as *non-analytic.*

Reply to the First Objection. The caveat in question can be disposed of in terms of the Second Reason for rejecting the appearance as an interpretation of the initiator that was given above. Again, if the concept of the object's affecting the mind were contained in the concept of the object's being given to us, such that the latter concept would not be logically independent of the former concept, Kant could not use the former concept to say, as he does in the passage in question, that the possibility of the object's being given to us *depends on* the object's affecting the mind.

19 See chapter 3 for an argument to this effect.

This dependence, and hence the logical *independence* of the concept of being given to us from the concept of affecting the mind is unaffected by the caveat that the object is given to us *through an intuition.* Consequently, the caveat can be included in the concept of an object's being given to us, such that the concept is understood as that of an object's being given to us *through an intuition*, and still the concept will be logically independent of the concept of an object's affecting the mind.

Before proceeding with a rejoinder to this reply, it should be made clear, for the purposes of our discussion, that it is agreed by both sides to this dispute that '*logical* independence' means that it *would not be a logical contradiction or inconsistent* for a proposition that contains one concept to be *affirmed* in the same context in which a proposition that contains the other concept is *denied.*

Rejoinder to the Reply. The reply doesn't understand the nature of the possibility in question. The reply takes the idea of a condition of a possibility of an object of a concept as it occurs at A19/B33 as a condition of which the concept in question is logically *independent*, whereas the passage can very readily be interpreted as using a notion of *logical* possibility. In that case, the concept whose possibility is in question would *not* be independent of the condition of the possibility. Consequently, the condition would be *entailed* by the concept. Accordingly, the passage could be read as asserting that it would be *logically* impossible for an object to be given to us independently of the object's affecting the mind.

The rejoinder could continue that the logical dependence just stated does not itself require that the dependence is reciprocal. That is, the entailment could go just one-way – there being no logical equivalence. Indeed, it is acknowledged that the converse entailment cannot obtain in this context, since it goes without saying that things in themselves can affect the mind, but everyone agrees that they cannot be given to us. To sum up, the concept of being given to us (through an intuition) can (and in the passage does) entail the concept of affecting the mind. So, affecting the mind is a necessary condition of the merely *logical* possibility that an object is given to us. Consequently, the reply notwithstanding, an object's being given to us is *not* logically independent of the condition that the object affects the mind. Therefore, the object that affects the mind *can* be the appearance.

Reply to the Rejoinder. If Kant meant the possibility in question to be merely logical, it would be difficult to makes sense of the qualification of the possibility that he introduces in the second edition. He there adds that affecting the mind is a condition of the possibility in question, "to man at least." The qualification allows that affecting the mind may not be a condition of the possibility for another sort of subject of intuition (to which an object can be given). Since the second edition explores the distinction between a sensible and an intellectual intuition in much greater detail than the first edition does, two possible interpretations suggest themselves. First, it may be that Kant is thinking of human intuition as only one among multiple possible sorts of *sensible* intuition. In that case, Kant may be intending to distinguish human intuition from other possible sorts of *sensible* intuition. Or, he may be thinking of human intuition as one sort of intuition *in general*, the other sort being intellectual, which belongs only to a "primordial being." In that case, human intuition would be said to be distinct only from a divine intuition.

In either case, however, the possibility in question is qualified. The qualification is actually a restriction, not to man alone, of course, but to a sort of subject of intuition that is identical to man. The question for the rejoinder then is whether it can provide a sense of merely *logical* possibility that is restricted to sorts of subjects of intuition that are identical to man. Since the rejoinder maintains that the concept of an object's being given entails that the object affects the mind, it would also have to hold that the *concept* of an object's being given is itself relative to subjects of intuition that are identical to human beings. Thus, not only would the nature of logical possibility be determined by human beings, but that of the concepts that logic orders and interrelates with one another would also be determined by human beings. So, the rejoinder carries the desideratum that it would need to explain such a notion of logic and of concept and provide grounds for attributing these notions to Kant in the passage in question. It is a large order to fill.

The Second Objection to the Argument. It might be further objected that changing the semantic character of the identity proposition from non-analytic to analytic does not adversely affect the value of the interpretation of the affecting object as the appearance, and hence as the object that is given to us. Consequently, the initiator can still be interpreted as the appearance. Whether the identity proposition is analytic or not should not affect the question of the correct interpretation of the object that affects the mind as it appears in the identity proposition. Rather, the latter question is independent of the former, and if the interpretation of the object that affects the mind as the appearance is correct, that should rather determine that the identity proposition is analytic, and the initial determination that it is not analytic should then be corrected. In other words, the argument has reversed the correct order of dependence between the semantic character of the identity proposition and the correct interpretation of the object that affects the mind.

Reply to the Second Objection. The reply is the same as that which was first given to the first objection. If the objection were correct, the condition that the object affects the mind could not be a condition that makes it possible for the object to be given to us, since the objection considers the condition to be part of the concept that depends on it, which would result in having it depend on itself, which would be absurd.

These three reasons are reason enough to give up on the currently popular version of the dual-aspect interpretation of transcendental idealism, or any other dual-aspect interpretation that takes the appearance as the initiator, as being able to provide a satisfactory interpretation of the initiator.

[9] The Dual-Object Interpretation of The External Objects That Get Our Knowledge Started. On this interpretation, in affecting our senses the external objects – things in themselves – *produce* an appearance in us.

To its credit, the dual-object interpretation is aware that if the *a priori* ways in which we are affected by external objects could be ways in which the objects exist prior to experience *in the order of time*, the ways themselves could not be *knowledge*, since all knowledge begins in time *with* experience. Consequently, for the sake of the possibility of *a priori* knowledge, the ways in which the objects, now things in themselves, exist cannot be the *a priori* ways in which we can be affected by the objects. As Kant so clearly says, the thing in itself is the "true correlate of sensibility" (A30/B45), which is one of the essential elements of *a priori* knowledge. Finally, since external objects continue to exist in their *own* ways even as they affects us, they must be manifested in our experience of the objects that *are* given to us, i. e. appearances. Appearances are therefore the *effects* of things in themselves.

Despite these advantages, the dual-object interpretation of the external objects that start our knowledge has four critical drawbacks. First, in taking the affecting objects and the objects that are given to us to be *different* objects (its being the dual-*object* interpretation, after all), it cannot make sense of something that counted in favor of the dual-*aspect* interpretation of the external object, namely, Kant's explicit statement that the affecting objects *are* the given objects.

A second, related problem is that in the *Critique*, an object's being *given* is logically independent of its being an *effect* of some object. Yet that independence is implicitly contradicted by the dual-object interpretation. On that interpretation, the (causal) explanation of appearances as *effects* of things in themselves (on us) *is* Kant's (logical) explanation of the possibility of their being *given* (to us). It is the misunderstanding that lead Strawson to declare Kant's transcendental idealism "incoherent."[20]

The third, again related, problem concerns our direct reference to objects. An object's being given to us makes it possible for us to *directly refer to* it, a reference Kant calls *intuition.* Indeed, it is only as a condition for intuition to "take place" that the object intuited is said by Kant to depend on its being *given* to us.[21] On the other hand, its being an *effect* of an object on us tells us nothing about the possibility of our direct reference's to it, or our intuition's of it, "taking place." So, as an interpretation of the external objects that start our knowledge, the dual-object interpretation commits an *elenchi:* The interpretation is quite beside the point of accounting for our direct reference to objects.

It should be said, however, that Kant's use of the notion of direct reference is not the same as that typically employed today in philosophy of language. Whereas Kant's use implies that a *sense* is associated with the reference of a representation or knowledge, somewhat in the manner that a Fregean sense is associated with a linguistic expression that is employed to designate an object, the typical use of the notion in contemporary philosophy of language eschews employment of such a sense, and instead explains the designation of an object by a linguistic expression in terms of something like a historical or a causal link between expression and object – to the possible exclusion of a way (sense) in which the object is given to the user(s) of the expression.[22]

The fourth problem is the thought that may lie behind Kant's statement that the object that affects us is the identical object that is given to us. Whereas an *effect* does indeed depend on *another* object as its *cause*, an object can depend only on *itself* for *it* to be *given* to us.

Strawson's dual-object interpretation exhibits both the advantages and drawbacks of the interpretation that I have given above. Strawson recognizes the necessary independence between the mind and the affecting object in Kant's system. And since he also is aware that for Kant the way in

---

20 P. F. Strawson, *op.cit.*, p. 16.

21 "takes place" is the translation of "*findet . . . statt*" at A 19/B 33 and "*stattfinden*" at A33/B49.

22 See, for example, Saul Kripke, *Naming and Necessity* (Cambridge, MA: Harvard University Press, 1980 [originally published 1972]), pp. 91–97, *et passim.*

which an object affects us is the way in which it is given to us, he rightly concludes that to make sense of "the affecting relation," the object that affects us must be understood independently of the way in which it is given to us. The way in question is, of course, (inclusively) either space or time. Since space and/or time necessarily belong to appearances (A38/B55), and given Strawson's recognition of the necessary independence between the mind and the affecting object in Kant's system, the affecting object cannot be understood as being spatial and/or temporal. Not only does this make it impossible to understand the affecting object as the appearance (since, again, the appearance must be understood as spatial and/or temporal), but it makes it impossible to understand what the affecting relation between the object and the mind might consist in, since the only way that we can make the relation intelligible to ourselves is a way that indeed is spatial-temporal.[23]

However, on my view, he nonetheless erroneously concludes that the object can only be the thing in itself.[24] He doesn't seem to recognize the synthetic character of the identity proposition (step 1) in the argument advanced in the previous section of this chapter, i. e. section 8. He thus burdens Kant with the double-object interpretation of transcendental idealism. He has unfortunately backed Kant into the disastrous view that we cannot *know* the objects that affect us, and *a fortiori*, cannot know them as appearances. For Strawson's Kant, the objects that affect us – things in themselves – are merely objects of thought, not anything we can intuit. (Contrary to Allison's methodological, dual-aspect interpretation of transcendental idealism considered in the previous section of this chapter, this last proposition does not entail that things in themselves are therefore not thought to exist, and are therefore not thought to be the sort of object that we can readily understand as affecting the mind.) Actually, it is close to what I think is the very opposite of what Kant was trying to achieve with his "affecting," and in that sense, "causal," theory of intuition. To reiterate, it is my view that in affecting us in a certain way – say, spatially – the object is given to us in that way. Finally, Strawson's criticism turns into ridicule when he explores the futility of trying to explain the very relation in which we are affected by things in themselves in terms that can be applied to experience – thereby satisfying Strawson's "principle of significance."[25]

---

23 P. F. Strawson, *op. cit.*, Part Four: The Metaphysics of Transcendental Idealism.
24 *op. cit.*, Part I, *et passim.*
25 *op. cit.*, p. 16, *et passim.*

[10] A Dual-Aspect Resolution of These Four Dual-Object Problems. Finally, we can try to resolve all four dual-object problems together. An object's affecting us in a certain way can tell us how we can directly refer to an object through an intuition – issues that are logically independent of one another – and hence how we can refer to a *given* object through an intuition, and therefore explain how our intuition of an object can "take place," only if the affecting object is *identical* with the object referred to – a condition that the dual-object interpretation cannot countenance. In other words, though the two questions – our being affected by an object (albeit in a certain way – through an intuition) and our direct reference to an object (its being given to us through an intuition) – are logically independent of one another, it is only an *identical* object that can be involved in their joint resolution. Consequently, we would have an explanation of how we can experience, and hence *intuit* (i. e. directly refer to), the object that affects us – the initiator. Obviously, even though the *thing in itself* affects us and in so doing *produces* an object given to us (the appearance), it is still impossible, as Kant repeatedly tells us, for us to *intuit* it.

Besides Strawson's dual-object interpretation of Kant's transcendental account of the possibility of our *a priori* knowledge of objects, the interpretation of the identical object just offered in the preceding paragraphs also constitutes a reply to McDowell's very similar (to Strawson's interpretation, that is) dual-object interpretation of Kant's transcendental philosophy. McDowell, on the view offered here, correctly says, "In experience we take in, through impacts on the senses, elements in a reality that is precisely not outside the sphere of thinkable content."[26] On the interpretation of the identical object offered here, the aspect of initiators, that they are *given to us* through intuition, readily qualifies as an "element" of which McDowell is speaking in this passage. Initiators *as given to us* readily fall within "the sphere of thinkable content," if the *objects* of such thought would be allowed by McDowell to be placed within the sphere thinkable content, even though Kant distinguishes in this instance between content (empirical intuition) and object (appearance).

However, the interpretation offered here also departs from McDowell's understanding of Kant, as it departs in exactly the same way from Strawson's understanding. Like Strawson, McDowell thinks that once "a reality" is taken from "outside thinking and judging" into the

26 Mc Dowell, John, *Mind and World* (Cambridge, MA, London: Harvard University Press, 1996), p. 41.

"thinkable content" of experience, as the previous paragraph explains, the only other reality to which Kant can have recourse in his account of how our knowledge gets started is the reality of things in themselves. And for McDowell this comes within a "whisker" of turning the success of Kant's account of how a reality provides a necessary constraint on thinking and judging (again, as described in the previous paragraph) into the disaster that the reality of the constraint is really "fraudulent by comparison" with that of things in themselves.[27]

Perhaps an example from early Russell might clarify the resolution being offered here of the various dual-object problems that have been highlighted by our discussion of Strawson and McDowell. Material objects *produce* sense-data, but these objects of our knowledge by *description* (the material objects) have nothing to do with the possibility that "colors, sounds, smells, hardnesses, roughnesses, and so on" (i. e. sense-data) can be the objects of our knowledge by *acquaintance.* Moreover, for Russell, material objects are themselves logical constructions from sense-data. So, whereas the objects that are the logical constructions from the sense-data *produce* the sense-data, our *acquaintance* with the sense-data is a condition that is logically prior to their being understood as the *effects* of material objects, which, from a logical point of view, are merely our own logical constructions from the sense-data in the first place.[28]

The point of the comparison with Russell is not that our interpretation of Kant's identical object somehow tracks Russell's distinction between knowledge by acquaintance and knowledge by description, which if it does, it does only erratically, nor that Kant's identical object in either of its aspects is anything like Russell's logical constructions, but rather that pointing to the causal origins of the *objects* of our intuition or acquaintance, appearances for Kant, sense-data – indeed, appearances *again* – for Russell, does nothing whatsoever to explain the possibility of our intuition of or our acquaintance with objects.

So, neither the currently popular dual-aspect interpretation of transcendental idealism nor the dual-object interpretation can provide a satisfactory interpretation of the initiator. Whereas the candidates of the

27 Ibid.

28 But see the body of the text above, pp. 8–9, which states that Kant's intuition should not be modeled on direct reference theory in philosophy of language, which itself is a descendant of Russell's theory of logically proper names, and thus of his theory of knowledge by acquaintance. Kripke's idea of an original dubbing of an object with a name belongs to Russell's legacy in this regard.

currently popular dual-aspect either cannot be thought to affect our senses without committing a *petitio* or cannot be thought to exist (because the affecting object is only a methodological construct without any ontological implications) and yet is supposed to affect our senses, that of the dual-object interpretation, though it can be thought to exist, cannot be given to us. Moreover, whereas the candidate of a dual-aspect interpretation that is *not* methodological, but that nonetheless takes the *appearance* as the initiator, still cannot adequately account for the non-analytic character of the identity proposition, the candidate of the dual-aspect interpretation just sketched can be *both* thought to exist *and* given to us.[29] This object is identical with the appearance insofar as it is the object that is given to us. In addition, it is also identical with the thing in itself insofar as it is the object that affects us. So, the initiator is now appearance, now thing in itself. We thus seem to have a promising candidate for a *dual-aspect* interpretation of the external object and, hence more generally, of transcendental idealism.

Summation. It was originally said that besides eliminating appearances as the initiators for Kant, a moment's consideration would also eliminate things in themselves from playing the same role. The neglected identity at A19/B33 makes it impossible for things in themselves, as well as appearances, to do the job. The eliminations is simple: As already said, things in themselves cannot be initiators because they cannot be given to us. Having precisely the opposite character from appearances, namely, that they are objects of thought alone, they can't possibly satisfy the condition that they are given to us (through an intuition). Whereas appearances cannot get our knowledge started because they cannot affect the mind, things in themselves cannot get it started because they cannot be given to us. These are the two necessary properties that are possessed only by the identical object itself – the initiator.

That is why I suggest that it is the *combination* of these two properties that for Kant is the initiator. My suggestion is that he has taken these two properties and elevated them into distinct and logically opposed concepts whose respective objects are *appearances* and *things in themselves.* This elevation, or reification, in turn unfortunately has led to the dual-object interpretation of transcendental idealism.

My view is that neither the dual-object nor any dual-aspect interpretation that takes the appearance as the initiator can do justice to the fun-

29 The appearance is not logically prior to experience (nor is it temporally prior, either).

damental identity proposition that is asserted at A19/B33 and upon which Kant builds the rest of his system. Only the recognition of the logically primitive nature of the concept of an initiator – an object about which it is just as necessary to ascribe the property that it is given to us as it is to hold that it affects us – can fend off the equally unsuccessful interpretations of dualism and of monism that have so far occupied most of our attention in the literature.

***

Before closing, it might be mentioned that this interpretation of the external object might seem similar to the double affection theory proposed by Adickes, who, confronting the famous dilemma posed by Jacobi, employed the same *traditional dichotomy* for interpreting transcendental idealism as that used by Jacobi in formulating the dilemma. According to Adickes, there is both an empirical affection, which is due to the appearance, and a transcendent affection, which is due to the thing in itself. On the view proposed in this chapter, however, the affection that accounts for our *direct reference* to the appearance is due, not to the appearance, but to the initiator that can be given to us as an appearance, and the affection that accounts for the appearance as an *effect* on us by an object is due to the thing in itself.

[11] Coda. Perhaps the popularity of the prevalent dual-aspect interpretation of transcendental idealism is a reaction to the dual-object interpretation of the initiator. Unfortunately, current dual-aspect commentators have confined themselves to the *traditional dichotomy* of transcendental idealism – appearance and thing in itself – to select their candidate for the external object.[30] This chapter, on the contrary, has tried to find a more promising dual-aspect interpretation by thinking outside the dichotomy. As a result, it has proposed a third concept of an object, one that contains the features of both *appearance* and *thing in itself*, which is just what Kant himself said it had to contain. Accordingly, one would get three interrelated *concepts* of an object – *the initiator* (the identical object that has been proposed since section 2 above), (*its*) *appearance* (one aspect of it), and *thing in itself* (the other aspect of it) (Bxx, Bxxvi, *et passim*).

30 See above section 6 and footnote 9.

Chapter 3 – A Criterion of Existence in General

[1] Introduction. It is doubtful that the previous chapter will actually end the controversy over whether Kant has a dual-object or a dual-aspect theory of reality. In fact, the dispute seems interminable. To review the dispute very briefly, is transcendental idealism committed to two types of object – appearance and thing in itself – or just one type, say, appearance, with the proviso that each appearance has two aspects – appearance, which is the true and serviceable aspect, and thing in itself, which is the methodologically prohibited one? In both the Preface and the previous chapter we have considered both sides of the issue, and have done so in regard to several of its facets. And at the conclusion of the last chapter, there was an attempt to resolve the dispute in favor of a dual-aspect theory, only a different one from the one that is currently popular.

Although the last chapter takes a view of the two aspects that is as epistemological as is the currently popular dual-aspect interpretation, it begins from a distinctly different origin from that which starts the popular version. The popular version begins with our *subjective conditions of knowledge* (in contradistinction with conditions inherent in objects *in themselves*) and only late into the investigation does it deal with the hoary problem of the nature of our relation to objects that results in our getting sensations from them – the so-called problem of affection. In contrast, the version of the dual-aspect interpretation offered in section 10 of the last chapter begins with the *objective* origins of our knowledge and asks, Which *objects* get our knowledge started? The answer is given in terms of the same relation of affection that enters the popular dual-aspect view only late in its discussion of the topic. It is that the identical object that both views are trying to determine is better understood as *the external object that gets our knowledge started.* It is argued that this object is none other than *the object that in affecting us is given to us.* In the course of reaching its conclusion, it rejects both the *appearance* and the *thing in itself* as acceptable ways of understanding the external object in question, i. e. the initiator.

Apart from whatever advance that may have been made so far toward the objective of resolving the controversy over the correct interpretation of transcendental idealism, including any ground that may have been covered in my previous book on Kant, perhaps the controversy would not be so intractable if progress might be made on another, relatively neglected, question – a question only skirted in my earlier work. *This is the question of whether a criterion of an existing object might be found in Kant's theory.* Since his *category* of existence is independent of the category of reality, we might expect that a *criterion* of the concept of an existing object, or, for short, a criterion of existence, would be independent of his concept of a real object.

If such a criterion can be found and if it can be applied to the external objects that get our knowledge started, the respective issues of these two chapters will be joined and the answer to the question of the criterion of existence can be used to reinforce our answer to the question of the external objects that get our knowledge started (the initiators). That is, our confidence in our interpretation of Kant's transcendental idealism in the previous chapter – that initiators are objects that in affecting us are given to us – will be increased to the extent that we can employ the criterion of existence in our interpretation of Kant's application of his concept of existence to his external objects that get our knowledge started. In a word, Kant can say of these external objects that they exist.

The question of the independence of a criterion of existence from a concept of a real object – of which Kant's category of reality is one such concept – raises the question of the independence of a criterion for the application or use of the concept of existence – necessary and sufficient conditions of its application – from the *meaning* of the concept, that is, from conditions that define the concept (its definition). I would submit that Kant himself was well aware of this distinction and employed it.[1]

Before we address that question, however, it should be made clear that the concept of existence *in general* the criterion of whose application or use is the topic of the present chapter is not the *category* of existence, and the criterion of the former that will be offered in the course of this chapter is not the criterion for the use of the latter that Kant gives in his discussion of the category. This distinction conforms to the primary objective of the book. Kant's *own* uses of the alethic modalities – *uses* that

1 It should thus be noted at this point that a *criterion* for the use or application of a concept is not at all meant to be a *sense* or *meaning* of the concept. A criterion here consists of conditions for using or applying a concept that has a sense or a meaning, but does not consist of conditions that, for example, an object must satisfy for a concept to be true of it. More about the distinction between meaning and criterion will be said later, in section 6 of this chapter below. The same distinction needs to be made later in the book, in chapter 5.

entail his idealistic compromise between existence and necessity, whose delineation is the primary objective of this book – belong to his *theory* of the possibility of *a priori* knowledge – his transcendental epistemology – a theory that only *mentions* the modalities as they are *used* in the *a priori* knowledge in question. His *own* uses of the modalities in his theory of the possibility of the knowledge do not belong to the knowledge, which itself, again, makes use of the modalities whose use the theory is supposed to explain. To repeat, it is the interpretation of Kant's *own* uses of the modalities, not their uses in the knowledge that his own uses are supposed to explain, that is the primary objective of the book in hand. This is not to say that it is the book's only objective. Chapter 9 in particular considers Kant's interpretation of our use of *de re* necessity in geometry and in causal judgments.

When in what follows Kant's *own* uses of the modalities are distinguished from their uses in the knowledge in question, his own uses can be found in such statements of his as:

> "Space is a *necessary a priori* representation . . . " (A24/B39, my emphasis); "this [*a priori* synthetic] knowledge is *possible* only on the assumption of a given mode of explaining the concept" (B40, my emphasis); "Time is a *necessary* representation that underlies all intuitions" (A31/B46, my emphasis); "it [i.e. the form of the intuition] belongs really and *necessarily* to the appearance of this object" (A38/B55, my emphasis); " . . . they [objects of sensible intuition] *must* . . . conform to the conditions which the understanding requires for the synthetic unity of thought . . . " (A90/B123, my emphasis); and "Nature, considered as nature in general, is dependent upon these categories as the original ground of its *necessary* conformity to law" (B165, my emphasis).

In addition, his uses of the modalities are also expressed in words other than 'possible', 'necessary', and 'must'. That is, he employs various words to express his uses of the modalities, not merely just certain words that we generally employ to express our uses of them. For example, in setting up the problem the Transcendental Deduction of the categories is supposed to solve, we could say that he is using the modality of *contingency* when he uses the words 'might be' to say,

> "Everything *might be* in such confusion that, for instance, in the series of appearances nothing presented itself which might yield a rule of synthesis and so answer to the concept of cause and effect" (A90/B123, my emphasis).

For the sake of clarity, it should be added that the distinction being drawn between Kant's own uses of the modalities and their uses in his expressions of the judgments belonging to the knowledge that he is talking about is *not* reducible to the distinction between the *use* and *mention* of an expression or of a concept. Nor is it reducible to the object-language/metalanguage distinction. For Kant *own* uses of the concept of existence occur in his talk about *objects* as well as concepts, and accordingly, his criterion for his uses of the concept determines his application of the concept to *objects*, not to concepts. Of course, these are objects in his system; but they are objects nonetheless – something that would be ruled out if the distinction presently being drawn in the interpretation of Kant were the *use-mention* distinction or the object-language/metalanguage distinction. The point being made can be readily confirmed by returning to the passages from Kant just above. It is clear that he is talking about objects in several of them.

Returning now to the general distinction between a criterion of a use of a concept and its meaning, it was stated above that the application of the general distinction to the concept of existence in general and its corresponding criterion can be based on the distinction Kant himself draws between the category of existence and its criterion of application or use. First, the categories in general are defined by Kant – or at least, according to one of his definitions of them – as "concepts of an object in general, through which the intuition of an object is regarded as determined in respect of one of the logical functions of judgment" (B128, my substitution of "through which" [Kant's *dadurch*] for Pluhar's "whereby" and for Kemp Smith's and Guyer-Wood's "by means of which," which otherwise is often used by translators to translate Kant's use of *vermittelst*). This is obviously a more complicated definition than the more common concepts of "objects of intuition in general" (A79/B105) or even more simply, concepts of "an object in general" (A247/B304, cf. A254/B309).

On the other hand, Kant introduces the *criterion* of the application or use of a category, as an additional determination of the category. Following what has come to be recognized as the first step of the B-Deduction, famously presaging Frege, Kant speaks of the necessity of an intuition "at hand" if a category is to be given "sense [*Sinn*] and meaning [Bedeutung]" (B149, my translation). Such an intuition is the core of the criterion in question. A category is determined further through a *schematism*, which involves the intuition of time, and then, finally, on top of that, through a principle. Given our own very limited interest in the distinction between the criteria for the application or use of the *categories* of the modalities and their meaning, suffice it to say that their criteria are found among the schematisms and the Postulates of Empirical Thought.

As has already been stated, however, the concept of existence in general and its criterion of application or use belong to Kant's *theory* of the categories and of our uses or applications of them in our *a priori* knowledge of objects. So, neither the modal categories themselves, their schematisms, nor the Postulates of Empirical Thought are presently of any use to us, except as providing a pattern we might follow in finding the concept in question, i. e. that of existence, and its criterion of use or application.

Nonetheless, despite the distinction between a concept of existence in general and any criterion of its application or use, if a criterion of existence could be found in Kant, it might be used to say something about the original controversy over his dualism. This chapter is a search for such a criterion (without implying that once found, it is the only criterion of existence that can be found in the theory), and it indeed concludes with a suggestion for using the criterion that is found, not to resolve the dispute about Kant's dualism, but rather to help explain its rather pesky endurance.

Finally, the criterion of existence that we will be seeking needs to be distinguished from both of the two criteria of Kantian ontological commitment that were discussed in the Preface – the logical one from my previous book and the epistemological one that I arrived at in the previous chapter. Whereas the two criteria of ontological commitment actually *use* what I consider to be Kant's concept of existence, the criterion for using the concept itself must be independent of any further criterion that actually uses it in determining Kant's ontological commitment. Otherwise, it couldn't function as a criterion for the use of the concept: The determination would be circular.

[2] Kant's Own Use of the Concept of a Given Object In General. The previous section gave one of Kant's less complicated definitions of the categories as concepts of objects in general. Since I have just argued that the concept of existence in general whose criterion is the goal of our present search is distinct from that of the category of existence, when I speak of Kant's *own* use of the concept of a given object in general I do not mean by the latter a generic term for the objects in general that constitute the extension of the category of existence. Accordingly, our search will be for the criterion of the application or use of the concept of existence in general that is not the criterion for the category of existence. In any case, Kant himself makes it abundantly clear what the criterion of the *category* of existence consists in, given his discussion of its schematism and his Postulates of Empirical Thought.

The reason we must first introduce the concept of a given object in general before looking for the criterion of existence in his system, which may already appear quite obvious and self-evident in any case, is that the concept of existence for which the latter is the criterion is the concept of the existence of a given *object.* So, we need to be clear about what it is in general to be a given object the criterion of whose existence is in question. The criterion of existence that will be offered, in section 4 below, will duly confirm the view that we should first search for Kant's concept of a given object in general, for both the criterion of existence and the concept of a given object in general depend on the crucial concept of *intuition.* To anticipate, given objects in general are objects of intuition, and the criterion to be offered states that existence in general, i. e. the existence of objects, will belong to *all and only those objects that can be given as objects through intuition.* Although the criterion of existence in general does not entail that all given objects exist (for example, space and time do not, since they are given in part through a synthesis of the understanding [B160a]), it is definitely related to the concept of a given object in general.

We have already considered Kant's *Stufenleiter* – the classification of representations in general – in section 3 of the previous chapter. There we found two types of representation – intuition and concept – that are said to be knowledge. A third type of representation, sensation, "relates solely to the subject as a modification of its state" (A320/B377). The German word translated as 'relates' is '*bezieht*', which in Kant's system I think can generally be translated as 'reference'. Mention of my preference for 'reference' suggests a brief word of explanation is in order. My previous book, *Kant's Theory,* emphasizes a critical distinction between Kant's uses of the terms '*Beziehung*' and '*Verhältnis*', the former I would translate as 'reference', and the latter as 'relation'. Roughly speaking, *relations* (*Verhältnisse*) obtain between or among representations or between representations and a faculty of representation or of knowledge, between representations and a subject to which they belong, and finally, between objects that are represented by representations that stand to each other in *relations.* References (*Beziehungen*), on the other hand, concern the relation that obtains between representations, knowledge, or a subject thereof and that to which they refer. That to which they refer is thus considered possible with respect to the representations, the knowledge, or the subject or one of its faculties. The two types of relation that our English word 'relation' translates can be connected in the following way: The reference (*Beziehung*) of a representation, knowledge, or a subject (S) to an object consists in a representation's (R) being related to S (*verhält sich zu*) in a certain way. Consequently, the relation signified by *Beziehung* is not identical with the relation signified by *Verhältnis*, though the former depends on the latter. *The reference* of an S to an object is *not the relation* (*Verhältnis*) in which S stands to a representation R; it (i.e. *Beziehung*) *is* rather S's *standing in* that relation (*Verhältnis*), i.e. its *being so related*, to R. To sum up, I would translate '*bezieht*' in the present passage from the *Stufenleiter* as 'refers to.' I am reluctant to call the subject's state of mind in the present case an object.

Returning to the passage itself, we find support for the view that Kant considers sensations of themselves to not refer to *objects*, let alone given objects – they are non-intentional. Perhaps the *locus classicus* of this view is Rolf George's *Kant's Sensationism.*[2] As the *Stufenleiter* states, sensations refer rather to a subjective state. Consequently, we can't look to them of themselves to give us references to given objects.

The categories cannot do it, either. For the categories refer only to objects *in general* (*überhaupt*). When Kant speaks of a manifold of intuition *in general* (*überhaupt*), he abstracts from the form of an intuition and is left with the categories. He does so with respect to space and time (B162–3) or appearances, i.e. nature, or experience (B165), and he says that as *x in general*, *x* lies *at the basis* of the *determination* of the manifold of intuition in *x*. The determination consists in the respective positions of the elements of the manifold being fixed in *x*. Thus, since space in general lies at the basis of a perception of a house, the manifold of its intuition is *determined*, or *fixed*, in space. The idea is that the parts of the house are spatially determined, or fixed, with respect to each other and to other things in space. The contingency that appearances of the house might be confused – the problem addressed by the Transcendental Deduction and referred to in the previous section – is bypassed. A similar analysis can be applied to the manifold of an intuition of water freezing and its determination in time that rests on the determination of a manifold of intuition in general, which just *is* the concept of cause, if the form of the intuition, time, is abstracted from the intuition. Perception of the freezing, since it depends on the determination, or irreversibility, of the stages of the freezing in time, must therefore conform to the concept of cause.[3] Appearances, or nature, must conform to law for the same reason – their manifold in space and time is determined on the

---

2 *Synthese,* 47 (1981), 229–55/

3 For a discussion of Kant's treatment of causality see chapter 9, sections 13–16.

basis of an underlying determination of intuition in general, that is, on the basis of the categories – nature in general. And the same analysis holds for experience. Perceptions have a determined or fixed place in experience because experience rests on the determination of intuition in general, that is, it rests on the categories. In that sense, experience must conform to the categories.

Finally, in all these cases, the determination of whatever manifold, be it of outward or inward intuition, appearances (nature), or experience (connected perceptions), consists in the *application* (*Anwendung*) of the synthetic unity of the manifold of a sensible intuition *in general* to the sensible intuition, and therefore consists in the *application* of the categories to our sensible intuition (B163–65). As just explained, this application is none other than *x in general* lying at the basis of the determination of the manifold in *x.* Kant considers the application an action of the *imagination*, namely, the understanding acting on sensibility according to the categories (§24). Altogether, the above is a brief explanation of his claim that the understanding prescribes its own *a priori* laws unto nature (B159, B164).

The point of this interpretation of text, however, is to get clear about Kant's notion of *objects in general.* They are intellectual abstractions that must be applied to sensibility or a manifold of intuition, if they are to determine any objects. But they themselves do not consist of any objects. A manifold of intuition or of objects or of perceptions may be arranged in a certain way in space, time, nature, or experience and that arrangement may be determined by an underlying sensible intuition in general, which consists of a category or of categories, if the form of the intuition, space or time, is abstracted from whatever intuition is involved. But the abstraction of sensibility leaves the representation of the synthetic unity of the manifold of a sensible intuition in general without an object – it leaves only a concept or concepts, i. e. a category or categories. So, though categories represent objects in general, they represent no particular objects – the 'objects' are only objects of the intellect.

The combination of sensations and the categories will be of no help. The combination will give us sensations in general, which would be difficult to understand in any case. We would be trying to determine sensations in respect of the logical functions of judgment. A function, Kant states, is "the unity of the act through which many representations are brought under one common representation" (A68/B93). In the case of sensations, the act would be a judgment in which sensations are brought under the categories. But categories are precisely the sort of concepts that *cannot* be applied to subjective states: They determine an order in which *objects* can be given. This incoherence is indistinguishable from that which attends the notion that sensations *in general*, say, a certain color in general, e.g., red in general. For the notion invites us to think of the determination of something *x* in red, as we are supposed to think of the determination of appearances in space and time as resting on space and time in general (i.e. the categories, once sensibility – the form of the intuition – is abstracted). But the idea of such an *x* seems absurd.

Perhaps Russell's sense-data determined as logical subjects of atomic propositions would approach the idea. But no categories have any such role to play in Russell's logical atomism. Thinking of sensations as sensations in general leaves out what is essential to Kant's notion – that in the present instance the *in general* signifies the conformity to the categories once the form of the intuition has been abstracted from the manifold. The problem with trying to fit Russell's logical atomism into Kant's system is not only the absence of any categories, but the absence of any form of intuition. That is why Kant would say that Russell's logical atomism lacks any particular objects, notwithstanding Russell's claim that sense-data are indeed particular objects.

The result of our analysis is that neither sensations, the categories, nor their combination, give us particular objects. The absence of intuition alone explains their absence. Of the three relevant types of representation from the *Stufenleiter* – sensation, concept, and intuition – intuition alone accounts for the representation of particular objects. The lesson for our search for a criterion of existence is that since it is the existence of objects that Kant is speaking of, the criterion must take intuition into account. This will be amply fulfilled when we finally formulate the criterion in section 4 below.

[3] Original Intuition and Divine Self-Consciousness. In several passages in the second (B) edition of the *Critique* Kant talks about the strange faculty of knowledge belonging to a "primordial," or "original," ("*ursprunglich*") or a "divine" ("*göttlichen*") being.[4] Though this cognitive power is unfamiliar to us, he still uses it in his attempt to put the limi-

4 Perhaps Kant switched from "primordial," or "original," (*ursprunglich*) to "divine" (*göttlichen*) when he came to characterizing the same being's cognitive self-consciousness because he wanted to reserve "original" for his characterization of *our* self-consciousness, that is, the self-consciousness of a being whose intuition is sensible. So he chose "*göttlichen*" instead for the self-consciousness of a being whose intuition is intellectual and "original" (B 142).

tations of our own faculties of knowledge into perspective (B72, B135, B138–39, B145).

The objects of a divine self-consciousness are said by Kant to exist through the self-consciousness itself (B139). This is hard for us to understand, since being familiar only with our own kind of self-consciousness, we think of the existence of its objects – indeed, the existence of any objects – as independent of any self-consciousness. So it cannot be because it is a self-consciousness that a divine self-consciousness is said to have the peculiar property that through itself its objects at the same time exist. Rather, it must be because a divine self-consciousness has a property that a self-consciousness in general does not have. Evidently aware of this, Kant explains that a divine self-consciousness has the unique property that through itself a "manifold" is given (B135, B138). He further stipulates that this would be a "manifold of intuition" (B138). Again, however, this is hard for us to understand, since being familiar only with our own kind of manifold of intuition – the sensible kind – we think of the existence of objects as independent of any manifold of intuition. So it cannot be because a manifold of intuition is given that a divine self-consciousness has the peculiar property that through itself its objects at the same time exist. Rather it must be because a manifold of an original intuition has a property that a manifold of intuition in general does not have.

We are supposed to conclude that whereas neither a self-consciousness nor a given manifold of intuition separately entails the existence of an object, they can do so together. But even this stipulation is hard for us to understand, since we are familiar with a *combination* of intuition and self-consciousness in mathematics that does not entail the existence of an object (B137–38, B147). We therefore think of the existence of objects as independent of a combination of a given manifold of intuition and a self-consciousness. So it cannot be because of a combination of a given manifold of intuition and a self-consciousness that a divine self-consciousness has the peculiar property that through itself its objects at the same time exist. In fact, that purported explanation is incoherent, since a combination of a given manifold of intuition and a self-consciousness is an action that can involve only a given manifold of intuition and a self-consciousness that are precisely *not* original or divine, respectively. Rather the explanation we are looking for must lie in a relation between a manifold of an original intuition and a divine self-consciousness that a combination of a manifold of a human intuition and a self-consciousness such as ours does *not* have.

Those familiar with Kant's stipulations on this score may immediately suggest that the relation in question is *equivalence.* More specifically, it is an equivalence between two mutually complementary ascriptions. A strict statement of the equivalence requires a brief explanation of certain technical terms, some of which have already been used above without explanation. *Combination* is a term for an *action* of the *understanding*, whereas *self-consciousness* is a term for the *representation* of the understanding; self-consciousness in turn is represented by '*I am*' (B138–39). A combination is *intellectual* in all and only contexts in which the subject is (self-)conscious of the combination; otherwise it is *imaginative* (B151, A78/B103). (Not all work of the imagination, however, is combination. Some is mere *association*, which is a *synthesis*, or a "putting together," among representations that can belong only to empirical, not *a priori* knowledge [A77/B 103, B152]). Now the equivalence can be stated strictly: *An understanding is intuitive if and only if the intuition is intellectual.* So, "Where is the problem?" it may be asked.

The problem is that the answer is no clearer than the question: How are we to understand an intuitive understanding or an intellectual intuition?[5] We are investigating the cognitive power of a divine being in order to find a criterion of an existing object in general, whether it is an object of divine or of human knowledge. The prospects of finding what we are looking for will hardly be improved so long as we try to plumb the depths of an equivalence between terms – intuition and intellect, or understanding – that for all we know ought to be independent of each other, in the sense that one can be true of a cognitive power in the absence of the other. Indeed, the method of explanation I have adopted in this chapter is the ordinary one of looking for a universal in given partic-

5 See Rae Langton, *Kantian Humility* (Oxford: Oxford University Press, 1998). Of this concept of intellectual intuition Langton says, "we strike depths which I have no ambition to plumb," p. 45. See also Ralph C.S. Walker, *Kant, The Arguments of the Philosophers* (London: Routledge & Kegan Paul, 1978). He also pronounces the concept a "mysterious idea," p. 30.

In his translation of the *Critique of Pure* Reason (Indianapolis/Cambridge: Hackett Publishing Company, Inc., 1996), p. 22, Werner S. Pluhar directs the reader to Kant's *Critique of Judgment*, § 77, where the concept is explained in terms of a discussion of teleology. At the same time, Pluhar also refers the reader to his own Introduction to his translation of the *Critique of Judgment* (Indianapolis: Hackett Publishing Company, 1987). But since neither Kant's nor Pluhar's discussion connects the concept directly to that of the existence of objects, which is our concern in this chapter, I will forgo dealing with those two discussions here.

ulars. In this case, the particulars include both divine and human cognitive powers. If such a universal can be found, it might then be used to go back to a divine cognitive power to understand it as an instance of the universal. That might make it easier for us to understand the possibility of an intellectual intuition or of an intuitive understanding.

[4] The Criterion of Existence Found, in Intuition. The intellectual character of an original intuition does have one advantage for our investigation, however. Since it makes a divine understanding and an original intuition equivalent, it allows us to proceed with our inquiry by considering a second edition passage that is framed in terms of intuition independently of self-consciousness. Understandably, this passage belongs to the Aesthetic. Like the first passage from the B-Deduction that we considered, but expressed a little differently, it stipulates that an original intuition "can of itself give the existence of its object" (B72).[6] The passage that we should recall in this connection stated somewhat differently that through a divine self-consciousness "the objects of [the self-consciousness] should at the same time exist" (B139).

If not the terms themselves of our previous arguments, the same form of the arguments can get us started. It is hard for us to understand Kant's stipulation of an original intuition as one that "can of itself give . . . the existence of its object," since being familiar only with our own kind of intuition, i.e. sensible intuition, we think of intuition as *dependent* on the existence of its object (B72).[7] So, it cannot be because it is an intuition that an original intuition is said to have the peculiar property that it "can of itself give . . . the existence of its object."

Now, however, our argument can depart from the old form and take on a new one. If we remove the unique property belonging to the original intuition, viz., that "*of itself*" the intuition can give the existence of its object, we can abstract a property that is *common* to original and human intuition. This would be the property that *through an intuition the existence of its object can be given.* This property, however, creates a problem. An existing object is an object of a sensible intuition only if the existence is already involved in its being given; that is what the dependence on the existence amounts to. To put the point another way, a sensible intuition does not even have an existing object unless one

6 This is supposed to be an entailment of the concept of the very faculty being mentioned – original, or intellectual, intuition – the faculty that commentators find so mysterious.

7 See note 2 above.

plays a part in its existence being given to the subject(s). For example, our outer intuition, which is spatial, has no existing object: "we can quite well think of [space] as empty of objects" (A24/B38–39).

Let us therefore qualify this property of an intuition. Let us say that an intuition, whether of itself or depending on the existence of an object, can give the existence of its object. And we shall mean by the dependence just mentioned that the existence plays a necessary part in its being given. This raises the question of which existing object among all existing objects is the object whose existence is being given. The answer is that it will be none other than the object whose existence the intuition in fact depends on – the existence that in fact plays the necessary part in an object's existence being given. This determination of a particular object indicates a necessary further qualification of the property in question. It will now be the property of an intuition, whether of itself or depending on the existence of the object whose existence it in fact depends on, that it can give the existence of its object.

Understood in terms of this property of an intuition as finally formulated, we can state our criterion of existence: *All and only objects that can be given as objects through intuition exist.* The difference between our sensible intuition and an original, intellectual intuition is that ours depends on the existence of the object, if the existence is to be given to the subject, whereas an original intuition does not have that dependence; rather, it is through the intuition "of itself" that the object's existence can be given.

[5] The Criterion and the Necessity of Transcendental Idealism. Before testing our criterion for adequacy, the difference between original and human intuition can make transcendental idealism especially compelling. *Let an intuition be such a representation that its determination belongs to an object if and only if the object is given through the intuition.* More specifically, if the object exists and the determination belongs to it, the existence of the object would be given through the intuition. For example, the form, or determination, of an outer empirical intuition, space, belongs to the corresponding appearance if and only if the existence of the appearance is given to us through the intuition.

It follows from the stipulation on intuition just given *that a determination, or form, of an intuition that depends on the existence of its object for the existence to be given cannot belong to the object.* For if it did, there could not be the dependence, since the existence of the object would *already* be given through the intuition. A determination that belongs to an object cannot also belong to an intuition that depends on the existence of the object for the existence to be given through the intuition, since that would have the intuition depend on itself, which is absurd.

To put the point another way, a form, or determination, that necessarily belongs to an object of intuition cannot also belong to an object whose existence is necessary for the existence to be given through the intuition. That would have the determination serving as its own necessary condition, which, again, is absurd. For example, since space or time necessarily belong to an appearance (A38/B55), it cannot belong to the object upon whose existence the intuition depends for the existence to be given to us (in the appearance).[8]

In case the intuition is not thus dependent on the existence of the object for the existence to be given, the determination in question would belong to the *object in itself.* That is precisely the nature of an original intuition. But because human intuition depends on the existence of its object for the existence to be given, the determination of the intuition cannot belong to the object in question – it can belong only to an *appearance* of the object. To do so, however, the dependence must be effective. And it is effective if and only if the intuition contains "[t]he effect of [the] object on the faculty of representation," namely, *sensation* (A19–20/B33–34). That makes the intuition *empirical.* Consequently, the object of an empirical intuition is only an *appearance* of the object (the *initiator*) on whose existence the intuition depends for the existence of that object to be given to us through the intuition.[9]

Obviously, the appearance cannot be the object upon whose existence the intuition depends. Rather, the converse is true: The appearance depends on the intuition for it to be given to us, since it can be given to us only through a determination that belongs to our intuition, i.e. space or time. But if the appearance depends on our intuition in that respect and our intuition in turn depends on the existence of the independently existing object for its existence to be given to us, it follows that an appearance depends on both our intuition, for it to be given to us as an

8 Both ways of making the point can be viewed as responses to a criticism of P. F. Strawson's. See his *op. cit.* The criticism is that Kant fails to argue for a "fundamental . . . complex premise of the Critique," p. 250. This is the premise that "any knowledge involving perception that is the outcome of our being affected by things existing independently of perception cannot be knowledge of the things as they are in themselves; it can be knowledge only of those things as they appear," ibid.

9 The necessity of sensation in the representation of existence will be explored further in the next chapter.

object, and the independently existing object, for the existence of that object to be given to us through the intuition, and thus through the appearance.

[6] Testing the Criterion for Adequacy. Now we can test for the adequacy of the criterion against several major concepts from the *Critique*. First, the criterion can dispel the air of mystery surrounding Kant's idea of a divine cognitive power through whose representation the object of the representation at the same time exists. Such an object is called by Kant a noumenon in the positive sense. If we do not find the criterion mysterious, we should not find the idea mysterious, either, since the object of the representation satisfies the criterion. But this result was foreordained by the very method that led to the criterion. The criterion was based on a property that was abstracted from a divine cognitive power as well as from a human one. So, the object of such a divine representation had to satisfy the criterion.

Let us next consider an object of our understanding that *cannot* be asserted to exist. This would be an object of our thought of *things in themselves* – what Kant calls a noumenon in the negative sense, since "it is *not an object of our sensible intuition*" (B307). Kant says that such a noumenon extends beyond appearances "*problematically*" and consequently cannot be the object of our *assertion* (A255/B310). Since the modal category that rests on the judgment-function of assertion is existence, a noumenon in the negative sense cannot be *asserted* to exist. This fits our criterion of existence, since it is based precisely on the absence of even a merely possible intuition of the object.

One might object that this result contradicts what was said in sections 8–10 of the last chapter about one of the shortcomings of the currently popular dual-aspect interpretation of transcendental idealism and, correlatively, one of the virtues of the dual-object interpretation. It was argued there that as an object of a thought that did not imply the existence of the object, the currently popular dual-aspect interpretation of the thing in itself could not account for Kant's thesis that we are affected by external objects, since any object taken to affect us must also be taken to exist. On the other hand, both the dual-object and my own interpretation of the thing in itself accords it the existence that is necessary to think of it as the object that produces appearances in us. Yet here it seems that I have just adopted the popular dual-aspect interpretation, since I have claimed that we cannot *assert* that Kant's noumena in the negative sense exist. This appears no better than saying that it is a mere *methodological* concept.

The objection is based on a confusion. My position is that Kant is quite consistent in maintaining both that we cannot intuit things in themselves and also that in affecting us they produce appearances in us. The latter thought clearly entails the thought that they exist. The former thought, however, entails that, if we accept the criterion of existence I have just offered, when Kant says that things in themselves produce appearances in us, the criterion for applying the concept of existence to an object is *not* satisfied. In short, on my interpretation of transcendental idealism, whereas the thought (and its expression) of the efficacy of things in themselves implies the thought that they exist, the criterion of existence remains not satisfied (because things in themselves cannot be given through intuition). On the popular dual-aspect interpretation, however, the first thought *cannot* imply that they exist, because the concept of things in themselves is only a methodological concept on that interpretation.

When Kant is adducing things in themselves as the causes of appearances in us the question of the satisfaction of the criterion of existence does not arise. For when speaking of the causes of appearances Kant is engaged in the theory of knowledge – epistemology – and not in the knowledge he is thereby explaining. Similarly, when he says that we cannot intuit things in themselves, and hence that we cannot assert their existence, he is still engaged in epistemology, and it is his view that we cannot know these objects that we think exist – these things in themselves – because we cannot intuit them. That the criterion of existence is not satisfied in this case doesn't keep him from saying that we cannot know these existing things; in fact, just the opposite is the case: It is precisely because the criterion is *not* satisfied that he can say of these objects that we can merely *think* their existence, but cannot know them. Otherwise, he would be in the unenviable position of being able to express thoughts of only those objects we can know, which would commit him to a form of verificationism – something we would not want to attribute to him, I believe.

The distinction I am drawing is between Kant's use of the concept of existence in his epistemology and the use of it in the knowledge under examination. The criterion of existence belongs to his epistemology, but it is not the epistemology that must satisfy the criterion; it is the knowledge being explained that must satisfy it. The epistemology does not contain intuition; it is the knowledge in question that must contain it. So, Kant can meaningfully think of things in themselves through the concept of existence – he can think of them as existing – in his epistemology, and recognize that the criterion of existence is not satisfied. According to my interpretation of Kant's criterion of existence, the concept of existence cannot be employed or applied to an object unless the object can be given through intuition. That is, it can only be applied to objects that can be given through intuition. But the possibility of its being given through intuition is a condition only of the use of the concept of existence or its application to an object. But this is no bar to our thinking of objects through the concept even if we cannot intuit them. The meaning of the concept is thus distinct from the conditions belonging to the criterion of its use or application to objects.

To sum up, the epistemology provides a criterion of the use or application of a concept to objects belonging to the knowledge the epistemology is supposed to explain; but there is no demand that the epistemology itself satisfy the criterion, since it makes no claim to knowledge such as those that are made by the knowledge in question. Contrary to Strawson's dictum that Kant repeatedly avows an empiricist "principle of significance," suggesting a verificationist or Wittgensteinian theory of meaning, I am suggesting that while the criterion of existence does indeed contain the requirement of intuition, which might meet such verificationist demands on claims of knowledge, its own formulation in epistemology need not itself be subject to such demands, and they are not so subject in Kant's epistemology.

Moving on, it might be thought that a counter-example to the criterion can be found in the manifold of our a priori sensible intuition, such as that of space or time. The manifold can be given through our intuition, and yet it cannot be said to exist. The reply is that the manifold cannot be given as an *object* through the intuition. Rather, the object would be something that can be given only through a *synthesis* of the manifold (B160a).[10] As such, it could belong to mathematics. As already noted, however, mathematical objects cannot be said to exist. Existing objects that have arithmetic forms can be given only through *empirical* intuition (B15). And the same is true for existing objects that have geometric forms (B156). This distinguishes them from the a priori objects of mathematics. As combinations of the understanding, the latter are "brought about" (B138)[11] only through a synthesis of the understanding and therefore cannot be said to exist.

10 I have discussed this issue at length in, *Kant's Theory*, pp. 219 ff.
11 Pluhar trans.

In the same vein, empty space and time cannot be thought to exist "without absurdities," Kant claims (B71). Yet empty space (and presumably empty time) can simply be thought (A24/B39). In a well-known footnote to the B-Deduction already referred to just above, Kant claims that space, "represented as *object*, (as we are required to do in geometry) . . . contains *combination* of the manifold" (B160a). As such it is a "*formal intuition*" and is empty. The same point is made about time. The note concludes that space and time are first given as intuitions through "a synthesis." Consequently, it is space and time as (empty) objects given through *synthesis*, not intuition, that cannot exist, which is precisely what our criterion would imply.

Objects of our a priori sensible intuition as determined by a *synthesis* of our *a priori*, or productive, imagination can be said to exist, but only upon further qualification. We have already seen that for the existence of objects of sensible intuition to be given the existence itself must be involved. We will deal with this dependence, and thus the existing objects of a priori intuition, i.e. appearances, at the end of this section.

Objects of what we ordinarily call imagination and Kant calls reproductive imagination are also the result of *synthesis – empirical synthesis.* They, too, cannot be said to exist. Their synthesis consists of an individual's mere associations, which are "entirely subject to empirical [psychological] laws . . . of association" (B152). A distinct *empirical synthesis* – an action of the *imagination* that has *a priori* cognitive value (B160b) but is itself empirical – is necessary for objects of perception. Kant calls this synthesis the *synthesis of apprehension* (B160). It is a synthesis of "the manifold in an empirical intuition" (ibid.). Since objects of perception are the results of this synthesis, however, we could not count them among existing objects over and above the objects we would have already counted as the objects of the empirical intuitions involved in the synthesis, namely, appearances, unless, that is, we are prepared to overpopulate the world of existing objects. The same objection would hold for objects of experience, which itself is a *synthesis* of perceptions.

Besides overpopulation, there would be the objection that it would be impossible to find any intuitions of the objects of perception or of experience. Empirical intuitions would have to have been preempted by appearances on the pain of an infinite regress of empirical intuitions – we would have to have empirical intuitions of the objects of perception as well as empirical intuitions of the appearances that are perceived – and *a priori* intuitions of themselves, we have already noted, cannot give the existence of their objects. Since those are the only two kinds of sensible intuition that we can have, Kant would have to confer existence on objects that we could not intuit. Kant would then be faced with the difficulty of distinguishing them from those other objects that are thought to exist even though they cannot satisfy the criterion of existence, viz., noumena in the negative sense. The same considerations would bear upon the objects of "all possible perceptions" (B165) – nature as the "sum of all appearances" (B163) – and also upon the objects of "experience in general" and correspondingly upon the objects of "*nature in general,*" which can tell us what can be known as an object of experience (B165).

Though a subject such as ourselves, apart from its own inner intuition, can be conscious of and determine its own existence, which entails that the subject can be said to exist, that does not constitute a valid objection to the criterion, since the subject can still be given (to itself) through that intuition (B157–58, B158a).

Finally, of course, there is the "undetermined object of an empirical intuition" – an appearance. Such an intuition contains sensation, which is the effect of the dependence of human intuition on the existence of its object that we have already discussed (A19–20/B33–34). Since an appearance can be given through intuition and can be said to exist – indeed, its existence is the existence that the existence of any other object of our knowledge must refer to – it, too, tends to confirm our criterion.

[7] The Criterion of Existence and the Controversy over Transcendental Idealism. The underlying criterion of existence we seem to have found in the Aesthetic and the Logic of the *Critique* is like Quine's quantificational criterion of ontological commitment (to be is to be a value of a bound variable)[12] – non-committal about the kind of object that can be said to exist. The difference between Quine's criterion of ontological commitment and Kant's criterion of existence is that, besides the distinction already drawn between the two types of criteria – one for ontological commitment, the other for the very use of the concept of existence in expressing a theory's ontological commitment – Quine's criterion is derived from *predicate logic* and thus the *logic* of *general* terms, whereas Kant's criterion of existence, in keeping with the epistemological thrust of the present book in contrast to *Kant's Theory,* provides a *perceptual* basis for our use of *singular* terms, the very terms Quine's criterion was supposed to help eliminate. So, neither the idea of concept (Quine) nor that of intu-

12 Among other places, see W. V. Quine, "Existence and Quantification," in *Ontological Relativity and Other Essays.*

ition (Kant) determines the kind of object that can satisfy their respective criteria of existence. That commitment is left to a specific theory to determine.

This neutrality of Kant's criterion may ironically have helped spawn the mutually incompatible interpretations of his concept of a *real* object and thus of his transcendental idealism – a controversy that has enjoyed such a vexatious endurance. On the one hand, the *univocal* conditions it stipulates may suggest that only a *single* set of conditions can determine the specific objects that can satisfy the criterion. For example, it may be argued that the conditions of a possible experience determine that only empirical objects can satisfy the criterion. Consequently, these would be the only objects covered by transcendental idealism, which attributes dual-aspects to them. On the other hand, the same univocal conditions stipulated by the criterion may suggest just the opposite – that one criterion can be employed for determining the ontological commitments of *any* particular theory that claims to be true. This is the use to which Quine put his criterion of ontological commitment. Since transcendental idealism posits two kinds of object – appearance and thing in itself – corresponding to two kinds of knowledge, human and divine, the criterion can be used to argue that Kant's dualism posits two kinds of reality.

If such speculation is plausible, our criterion of existence might at least partially explain the discouraging longevity of a dispute in the Kant literature that has no end in sight.

Chapter 4 – Sensation and Existence

[1] The Criterion of Our Sensible Intuition of an Existing Object. The criterion of existence *in general* given in the previous chapter abstracts from the distinctive characteristic that makes an intuition *sensible*, namely, that the intuition depends on the existence of its object, *if the existence is to be given through the intuition.* Therefore, the existence of the object is *independent* of the intuition in that sense. The defining characteristic of an *intellectual* intuition, on the other hand, is its independence, in the same sense, from any such existence of the object – the intuition "of itself" *gives the existence of its object.* Therefore, the existence of its object is *dependent* on the intuition, in the sense in question.

Obviously, no intuition is an intuition *in general*, that is, no particular intuition is an intuition *in general*; accordingly, no particular object is an object *in general*; and, in the same vein, no particular existence is an existence *in general.* The criterion of existence *in general* given in section 4 of the previous chapter determines the application of the concept of existence to objects that are intuited or to objects of intuitions. Intuitions belong to either of two sorts of intuition, intellectual or sensible. The existence of their objects is subsumed under a general concept of existence whose application to the objects is determined according to the criterion of existence *in general.* We thus have a general concept of existence and a criterion of existence in general that determines the application of the concept to objects.

Whereas the specific concept of an intellectual intuition cannot logically divide into (further) *sorts* of intellectual intuition the way that the general concept of an intuition divides (between intellectual and sensible), although it of course divides among *particular* intellectual intuitions, not only is the opposite true of the concept of a *sensible* intuition, but the latter *must* so divide. The former cannot divide because there cannot be different conditions under which the existence of objects is given through intuitions "of themselves." An intellectual intuition is precisely one the existence of whose object is given independently of *any* condition. That is what is meant by saying that the existence of the object is given through the intuition "of itself." Sensible intuitions, on the other hand, are precisely those intuitions that *do* depend on a condition, if the existence of their objects is to be given. The dependence on such a condition is part of the sense of the concept of a sensible intuition. The condition in question is obviously the existence of the object. As already concluded in the opening paragraph of this chapter, that existence must therefore be *independent* of the intuition: Since the existence's *being given through the intuition* depends on the existence itself, the existence must be independent of the intuition, that is, if the expression of the dependence ("depends on") is to make sense. And, again as already noted, this is precisely the opposite of the existence of an object of an intellectual intuition, where the existence is *not* independent of the intuition, in the sense in which the existence is to be given through the intuition.

The conditional nature of the relation between a sensible intuition and the existence of its object raises the question of the nature of the condition: Under what condition can the existence of its object be given through the intuition? That is, what does the dependence of the intuition on the existence of its object consist in? To be informative, any answer to the question must be independent of the concept of the intuition. Therefore, it must be possible that there can be more than one possible condition. Otherwise, any condition would be part of the sense of the concept of a sensible intuition; that is, if there *logically* could be only one condition, it would not be distinct from any sensible intuition, and thus could not possibly be a condition under which the existence of the object can be given through the intuition. Again, the idea of the *dependence* of the intuition on the existence of its object if the existence of its object is to be given would be absurd. The concept of such a condition must therefore be *general.* So, the very concept of the condition is the concept of *such* a condition, and the generality of the concept entails that there can be more than one condition that can fall under the concept. In other words, the representation of the condition is essentially conceptual or predicative.

There is an independent argument that can demonstrate the logical possibility of more than a unique condition under which the existence of its object can be given through a sensible intuition. The argument is that the categories require the possibility. The categories must be independent of sensibility, if they are to determine the intuitions and/or their objects, that come from sensibility.

To develop the point, we must mention that the categories' determination of intuitions consists in their assignment of the intuitions to particular logical functions of judgment (B128, B143). And here a word about a preferred translation of text is essential. In the so-called "clue"

(*Leitfaden*) to the discovery of the categories, Kant says, first, that they *extend over* (*auf . . . gehen*), objects, and then, a moment later, he says that they *extend over* (*auf . . . gehen*) objects of intuition in general (A79/B105). However, in the latter instance, Kemp Smith, Pluhar, and Guyer-Wood all translate Kant's "*auf . . . gehen*" as "apply to;" and in the former instance, whereas Kemp Smith and Pluhar also translate "*auf . . . gehen*" as "apply to," Guyer-Wood uses "pertains to," which does not conflict with the point that is about to be made. The point is that the categories *apply to particular* objects of a particular *sort* of, not just intuition in general, whether sensible or intellectual, but specifically *sensible* intuition. Kant expresses the idea of this application with his use of "*auf . . . angewandt.*" The only particular sort of sensible intuition Kant can specify, of course, is *our* human sensible intuition. The point is borne out in §§ 24 and 26 of the B-Deduction of the categories. To reiterate, the categories are said to *apply to* particular objects of particular intuitions belonging to specific sorts of sensible intuition, the one sort that can be determined being our human intuition, whereas they *extend over* objects of a sensible intuition *in general.* That is the difference that allows them to *determine* our particular intuitions, belonging, of course, to our sort of sensible intuition – human intuition.

Were the categories to *extend over only* objects of our human intuition, they would depend on the objects, in the sense in which a concept depends on its extension – which would be exactly the opposite of what the categories require if they are to determine intuitions and their objects in respect of the logical functions of judgment. They must be higher-level concepts than those they are supposed to comprehend, viz., the concepts of the objects of our intuitions (A69/B93–94), if they are to determine the latter in the respect in question. Consequently, the categories must possibly extend over objects of sensible intuition that can be given independently of sensibility. Since sensibility is the source of sensible intuitions through which the existence of objects that affect the mind in a certain way can be given, the categories must be independent of the relation of affection, a proposition that in any case is uncontroversial, regardless of any exception that may be taken to the argument that has just led to it. Consequently, Kant must allow for the possibility of a species of sensible intuition besides ours, that is, besides one whose members can only represent the existence of objects that can affect the mind in a certain way.

Kant himself bears out the necessary division of the notion of a sensible intuition. At the very beginning of the body of the *Critique* in its second edition he states that the relation of affection is a condition "to man at least" of the possibility of an object's being given to us (A19/B33). It is clear from both the Aesthetic and the Analytic that the notion of *sensible*, as distinct from the notion of *intellectual*, does not entail the notion of sensibility, and thus does not entail the notion of affection. The latter is a constraint on sensible intuition that is specific to our human intuition.

Since the concept of a sensible intuition must divide according to at least two sorts of possible intuition, the concept as divided is expressed as a sensible intuition *in general*. This follows the pattern in which *an intuition in general* expresses the division between two sorts of intuition, intellectual and sensible. And just as there is no intuition that is an intuition in general, that is, there is no *particular* intuition that is an intuition in general, there is no intuition that is a sensible intuition in general, and thus no particular intuition that is a sensible intuition in general. All intuitions, that is, all particular intuitions, that are sensible intuitions will belong to one sort of possible sensible intuition or another.

The same point holds with respect to the *objects* of sensible intuition and the *existence* of these objects, just as it did above with respect to the objects of intuition generally and the existence of those objects. Just as there is a sense of the concept of an existence in general whose application to particular objects of particular intuitions (belonging to specific sorts of possible intuition, intellectual or sensible) is determined by a criterion of existence in general, so there is a sense of the concept of existence of objects of a *sensible* intuition in general whose application to particular objects of sensible intuitions (belonging to specific *sorts* of possible sensible intuition, involving the relation of affection or otherwise) is determined by a corresponding criterion of the existence of objects of a sensible intuition in general. That is, there is a single criterion for the application of the concept of existence to objects of sensible intuitions regardless of whether the giving of the existence depends on the relation of affection or not. It would thus be expressed as a criterion of the existence of an object of a sensible intuition in general, only now the intuition that figured in the expression of the criterion would be represented as *sensible*. We would thus get the criterion: *All and only objects that can be given as objects through* a ***sensible intuition*** *exist.* This would be a criterion for the application of the concept of existence to objects whose existence is independent of intuition, that is, objects that have an *independent* existence, in the sense of *independence* in which the existence does not depend on the intuition if it is to be, but the intuition does depend on the existence, if the existence is to be given through the intuition.

If we now further determine the application of the concept of an intuition, we would take a specific sort of sensible intuition whose independently existing objects would correspond to the intuition. Accordingly, Kant divides such objects into those that affect sensibility and others. He speaks of a given object of an intuition as an object that affects the mind in a certain way, in "man at least" (A19/B33). This suggests the possibility that a given object of an intuition might not affect the mind in a certain way, albeit not a human intuition. However, it could be that Kant is alluding to an intellectual intuition; but since he has already drawn a contrast between intuition and thought, and since an intellectual intuition precludes the possibility of thought, it is likely that Kant has in mind the possibility of another sort of sensible intuition besides ours. So it seems that the further determination of the application of the concept of the existence of an object of which there is a sensible intuition is that the concept can be applied to an independent object that *affects* us.

The penultimate remaining question is the sort of sensible intuition that is an intuition of an independent object that *affects* us. Since *sensations* are the effects on us insofar as we are affected by objects (A20/B34), and since they are also the representations that constitute a representation's being *empirical* (A50/B74), a sensible intuition of an independent object that affects us must be an *empirical intuition.* Since, as we found in chapter 2, the external objects that get our knowledge started (the initiators) are the objects that in affecting us are given to us, the criterion of the existence of the initiators must be: *All and only objects that can be given as objects through* ***empirical intuitions*** *exist.* Modifying the *'objects'* as *'external' would* give us the following logical consequence of the criterion as just modified: *All and only external objects that can be given to us as objects through empirical intuitions exist.* Consequently, our human ***a priori*** sensible intuitions cannot represent these objects as existing. It might be noted that an intuition of ours that is *a priori* shares with an *intellectual* intuition the characteristic that it is independent from the existence of any object; but that is only because an *a priori* sensible intuition simply cannot refer to an existing object in the first place: A sensible intuition of ours must be *empirical* to do that. An intellectual intuition, on the other hand, does refer to an existing object, because it does so "of itself," and that is the sense in which it is independent from the existence of any object.

The last question is the sort of empirical intuition that is an intuition of an independent object that in affecting us is *given* to us. Since empirical intuitions contain sensations and sensations refer to states of the subject (A320/B377) or, alternatively, are effects of objects on the faculty of representation insofar as it is affected by them (A20/B34), the concept of such a subject – one such as ourselves – provides a principle for collecting diverse empirical intuitions into a particular class of empirical intuitions. These will be all the empirical intuitions belonging to a particular individual such as ourselves. If we then abstract from such a principle of collection, as we have abstracted from sorts of sensible intuitions to get a sensible intuition *in general*, and thus get objects of a sensible intuition *in general*, and as we had antecedently abstracted from the two sorts of intuition – sensible and intellectual – to get an intuition *in general*, which gave us our criterion of existence in general, we will get an empirical intuition *in general*, and an object of an empirical intuition *in general*.

Following the pattern of an intuition in general and a sensible intuition in general, no intuition is an empirical intuition *in general*. It would be a logical mistake to suppose that there could be a particular intuition of an empirical intuition in general. Accordingly, it would also be a logical mistake to think of there being a particular object that would be an instance of an object of an empirical intuition in general. Particular empirical intuitions and particular objects of empirical intuitions – appearances – are instances only of the *sort* of intuition or concept *empirical intuition* and the *sort* of object or concept of *appearance*. An empirical intuition in general is an abstraction from every such sort, and hence there can be no intuition that is an instance of such an abstraction.

The sort of empirical intuition we found in the paragraph before the last is the sort in question, and it consists of all and only those empirical intuitions belonging to a particular individual subject such as ourselves. In less technical terms, the principle is that of an individual human being (cf. A19/B33). The object of a member of this sort of intuition will be a particular appearance. It will be the object of an empirical intuition belonging to a particular subject such as ourselves, that is, a particular human being. This will be the subject (or the mind) that is affected by the external object that, as it is given to the subject as an appearance, starts the subject's knowledge.

It should be stated that no such intuition or its object is thereby relegated to the private domain of the subject. Even though the sensation involved is a subjective state, it obtains reference to an object as part of the content of an intuition whose form is common to all subjects that are affected by external objects that are given to them. These forms are the *a priori* intuitions of space and time. As such, they are necessary and universal for us and consequently are objective determinations of initiators.

[2] Confirmation of This Criterion and Kant's Idealism. The criterion for the application of the concept of existence to an independent object that affects us, viz., that we can empirically intuit the object, fits the interpretation of Kant's transcendental idealism that was given in section 10 of chapter 2. To repeat, the initiator is the object that in affecting us is given to us, and our dual-aspect/single-object interpretation of the idealism is that the identical object in question is this external object, that is, the one that in affecting us is given to us. The immediately preceding section has found, moreover, that the criterion for the application of the concept of existence to this object is the empirical intuition. Since the object is central to the interpretation of transcendental idealism offered by the present account, the criterion of its existence – empirical intuition – should tell us something about transcendental idealism as well. That is what this section of this chapter will try to do.

In the *Critique* Kant makes a statement about his transcendental idealism that, though it does not seem at all concerned with sensation, upon analysis leads to a confirmation of the criterion just given. And it does so by implying something important about the role of sensation in the idealism. The statement is: ". . . no determinations, whether absolute or relative, can be intuited prior to the existence of the things to which they belong, and none, therefore, can be intuited *a priori*" (A26/ B42). The determination under discussion here is space and the priority in question is obviously logical, not temporal. The statement thus says that space cannot be both intuited *a priori* and belong to things prior to their existence. If we interpret Kant's use of "things" ("*Dinge*") according to the epistemological interpretation that is presently being proposed, and as distinguished from the logical interpretation of my previous book, Kant is saying that space cannot be both intuited *a priori* and belong to initiators prior to their existence.

In addition, the statement also clearly implies a simplification of its meaning. That is, it implies that space cannot be both intuited *a priori* and belong to initiators, *simpliciter.* But for the moment I would like to ignore that implication and instead, for the sake of bringing out a certain feature of the role of sensation in Kant's idealism, deliberately misconstrue the statement as saying, more narrowly, that if space is intuited *a priori* (which for Kant is the *only* way space can be intuited), it cannot belong to initiators prior to *their existence;* but (and here is the misconstrual) it *can* belong to them *provided they exist.* In other words, the miscontrual consists in erroneously logically separating "intuited *a priori*" from "prior to . . . existence."

Given the misconstrual, we can ask, when could space be *both* intuited *a priori and* (contrary to Kant's actual meaning) belong to initiators, but *not* prior to their existence? The trivial, though incorrect, answer is: *when* initiators *exist in space.* But for Kant initiators exist in space only as *appearances;* that is, only as intuited by us, or as *mere representations* of ours. So, if space is intuited *a priori,* it does *not* to belong to initiators after all, as the obvious reading of Kant's statement implies. Rather, it can belong *only* to our *representations* of them. A bit later in the Aesthetic, speaking of either time or space, Kant says, "it belongs really and necessarily to the appearance of this object" (A 38/B 55). Since appearances are mere representations of initiators, it follows that, again, space and time belong only to our representations of initiators.

So the obvious reading of the statement has been the correct reading all along. Since space is intuited *a priori,* it cannot belong to initiators. But the incorrect reading – that it can belong to initiators in case they exist – has lead us to a major thesis of Kant's idealism, namely, that space can belong to initiators *only* as they are represented by us, or only as they appear to us who intuit them.

This analysis leads to a neat dichotomy that in turn will immediately suggest a particular role for sensation in Kant's idealism. An intuition of ours is either *a priori* or empirical. If it is *a priori,* it is a *determination,* space or time, not an initiator, that is intuited. If it is empirical, however, an initiator *is* intuited. One half of Kant's idealism is then made explicit when we consider *both* types of intuition. For then we see that since space or time cannot belong to initiators if either is intuited, the only objects to which either *can* belong are *appearances* of initiators, that is, the *representations* of initiators in our empirical intuitions of them. It is therefore only *appearances* of initiators that are determined by space or time and not initiators *in themselves* (since *'in themselves'* refers to a type of *determination* of initiators, i.e. apart from sensibility, the other type being *appearance,* i.e. in relation to sensibility).

This raises the question of when do initiators exist in space, and the answer Kant gives is that they do so when and only when they are *given to us* (as *appearances*). This for him entails that we are *affected* by the initiators (A19/B34). And we are affected by them only if they have an effect on us, which he considers *sensation* (A19–20/B34), such as "impenetrability, hardness, color, etc." (A 21/B 35). Therefore, initiators *would not exist in space unless we had sensations of them* (which would belong to our intuitions of them).

[3] Scientific Realism vs. Transcendental Realism. My account of Kant's notion of appearance is thus at variance with one put forward by Margaret Wilson.[1] She gave it a contemporary reading, taking it to be a material object that has only Lockean primary qualities, but no secondary qualities. Wilson's Kantian appearances would thus be congenial to contemporary scientific realists. Not that Strawson would endorse her understanding of Kant's appearances, but he has drawn a similar distinction between phenomenally propertied material objects – material objects with qualities as *we experience them* and objects *of* theoretical physics – the objects of scientific realism.[2] The point I am making about Kant, however, is that Margaret Wilson cannot be right about Kant's transcendental idealism, since Kant argues that initiators exist in space and time *only if* we have sensations of them. Kant duly acknowledges the scientific realists of his day when he allows that those realists would indeed have a conception of *things* (initiators) *in themselves* as material objects without secondary qualities (not *appearances* without secondary qualities, as Wilson reads Kant). But Kant makes that acknowledgment in a context in which things (initiators) in themselves are understood according to what he calls "the empirical understanding" (A29/B45). And that is precisely the type of understanding Strawson employs in drawing his distinction between phenomenally propertied material objects and the objects of theoretical physics. But the understanding that Kant employs is exactly the opposite of the empirical understanding; it is rather the *transcendental* understanding that gives us the transcendental concept of appearances. According to the transcendental understanding, initiators exist in space and time only if we have sensations of them (and at that they are mere appearances, not initiators in themselves).

[4] Guyer's Alternative Account of Transcendental Idealism. The interpretation of sensation as a *necessary* condition of the existence of initiators in space and time in my interpretation of Kant's idealism stands in contrast to the interpretation of sensation as a *sufficient* condition of the same existence of initiators in space and time in interpretations of others. This contrast sets off my interpretation of Kant's transcendental idealism

1 Margaret D. Wilson, "The 'Phenomenalisms' of Berkeley and Kant," in *Self and Nature in Kant's Philosophy*, Allen W. Wood, ed. (Ithaca: Cornell University Press, 1984).

2 P. F. Strawson, *Skepticism and Naturalism: Some Varieties* (New York: Columbia University Press, 1985), p. 42 ff.

(at least in regard to space) not just from Margaret Wilson's, but, for example, from Paul Guyer's as well.[3]

Guyer argues that for Kant Euclidean space is *necessary,* if we are to perceive an object, where I take Guyer to understand Kant, correctly in my view, to include sensation in perception. Consequently, in his interpretation of Kant's idealism Guyer must be taking sensation as part of a *sufficient* condition (i. e. perception) for the existence of initiators in space. Guyer then goes on to distinguish between two types of necessity pertaining to the spatiality of objects about which from the start of his book he has maintained Kant is confusedly ambivalent. On the one hand, there is the absolute type of necessity, which Guyer argues leads directly to Kant's transcendental idealism, and there is the conditional type of necessity, which is free of the implication of idealism.[4] Guyer criticizes and rejects the implication of absolute necessity that he claims to have found in Kant's theory of the possibility of our *a priori* knowledge of objects. Then, following the lead of Strawson, he embraces Kant's alleged use of a merely conditional necessity. This latter is the only necessity required in a transcendental theory of the possibility of *experience,* in contrast to a theory of the possibility of *a priori* knowledge. Consequently, a transcendental theory of experience has no need of Kant's transcendental idealism.

But if I am right, that sensation for Kant is a *necessary* condition for initiators to exist in space (and in time, I would add), we can dispense with Guyer's use of the distinction between the absolute and conditional necessity of Euclidean space in our perception of objects in our account of Kant's transcendental idealism, at least until we factor in the role of sensation as a *necessary* condition of the existence of initiators in space. According to my interpretation, any account of Kant's idealism would be incomplete unless it recognized this particular necessary condition in Kant's explanation. With regard to the role of sensation, therefore, my interpretation thus gives a more phenomenalistic (and thus less realistic) cast to Kant's transcendental idealism. In this respect, Guyer's interpretation is more like Strawson's, and mine is more like Jonathan Bennett's.[5]

---

3 Paul Guyer, *Kant and the Claims of Knowledge* (Cambridge, UK: Cambridge University Press, 1987) p. 364

4 Ibid.

5 Jonathan Bennett, *Kant's Analytic* (Cambridge, U.K.: Cambridge University Press, 1966).

The transcendental concept of things (initiators) in themselves for Kant is a concept of things (initiators) as they exist *apart* from space and time, where that concept concerns the *determinations* that belong to things (initiators) (going back to Kant's original statement [at A26/B42]). Since our having sensations is necessary for initiators to exist in space and time, then, if initiators exist apart from our having sensations, they also exist apart from space and time, and thus exist as initiators in themselves. To repeat, the opposing notions of appearance and initiator in itself arise in the first place only if we are inquiring about the *determinations* that belong to initiators, instead of inquiring whether there is either one or two types of *object* in the world (i. e. dual-aspect interpretations versus dual-object interpretations of Kant's transcendental idealism, e. g. Allison's versus Strawson's interpretations, respectively). I am not claiming that existing apart from sensations is the *only* way that there are initiators in themselves for Kant; but it is the way that stands in contrast to the scientific realists' conception of material objects that is discussed by both Wilson (in regard to Kant) and by Strawson (largely independently of Kant), a conception Kant himself assigns to the *empirical* understanding instead of to the transcendental understanding, which is the type of understanding Kant says he himself is employing.

Chapter 5 – Presupposition and Existence

[1] Further Differences Between Initiators and Appearances. In chapter 2, section 8, we found that whereas initiators exist independently of our intuitions of them, appearances of these same objects do not enjoy that independence. Consequently, initiators cannot be understood as appearances. Though appearances, like initiators, can be given to us, that same section argues that we cannot understand how appearances can so affect us that they can get our knowledge started; yet initiators are defined exactly that way.

This difference between appearances' dependence on our intuitions of initiators and the initiators' own independence of our intuitions of themselves is manifested grammatically. Although it is argued in section 2 of the last chapter that our specifically human criterion of the existence of initiators requires that our intuitions of them are empirical, it does not follow that initiators are the objects *of* those intuitions, since the latter are none other than appearances. Again, since initiators cannot be understood as appearances, and since the objects that we empirically intuit are initiators, the objects that we empirically intuit cannot be appearances, even though appearances are the objects *of* the same intuitions. Grammatically, this is a difference in case, between the accusative, or direct object, case ('we empirically intuit initiators'), and the genitive, or possessive case ('appearances are objects *of* our empirical intuitions').[1]

The difference can also be expressed in terms that render appearances *relative* to our intuitions of initiators and that keep initiators independent of the relation. Whereas appearances must be thought as relative to our intuitions of initiators, initiators cannot coherently be thought that way. For the claim of relativity to express a genuine condition on appearances, initiators themselves cannot be considered relative to our intuitions of *initiators*, since that would render the claim of relativity vacuous. In-

1 From this point on in the book, the use of quotations will become especially prominent. The following rule will be generally observed: Double quotes will be used if an expression is attributed to a particular person or text or if verbatim speech is being referred to. Otherwise, single quotes will be used, as, for example, if reference is being made to a common expression – a word, a sentence, or more.

tuitions of initiators would lack the independence from the initiators themselves that would be required for the claim to advance our understanding; that is, if initiators themselves were taken to be the objects of our intuitions of initiators, the claim that the objects of the intuitions of initiators are relative to our intuitions of initiators would make no sense at all.

Section 2 of the previous chapter states a further salient difference between initiators and appearances. It is that space and time can belong only to the appearances, if space and time are intuited "prior to the existence" of the initiators, that is, intuited *a priori.* How are we to understand this difference, that space and time can belong only to appearances? Assuming the intelligibility of this first question, a related question is, how can the existence of initiators in space and time coincide with the existence of appearances in space and time? Another way of putting the question is, how, on the one hand, can initiators exist in space and time, and yet, on the other hand, space and time not belong to them, if space and time are intuited prior to the existence of the initiators? A yet further related question is, how can *both* initiators in space and time *and* their appearances in space and time *depend on* our intuitions of the initiators? These three, related questions can be answered as follows.

In answer to the first two questions, section 2 of the last chapter argues that since initiators exist in space and time only as appearances, and since space and time necessarily belong to appearances, it is a logical truth that initiators and appearances jointly exist in space and time, if either one does. That is, the existence of one in space and time is equivalent to the existence of the other in space and time. To make the point another way: Appearances just are initiators insofar as the latter exist in space and time. On the other hand, if space and time themselves are intuited, and hence intuited *a priori*, the intuition is not empirical. In that case, since initiators are intuited only empirically, initiators are *not* intuited, and therefore do not exist in space and time. Consequently, in that case, space and time do not belong to them. Given the equivalence of the existence of the two in space and time, if space and time are intuited *a priori*, they don't belong to appearances either. Finally, since space and time *necessarily* belong to appearances, it follows that if space and time are themselves intuited (*a priori*), appearances themselves don't even *exist.* (In that case, space and time would be empty; that is, no objects – including initiators – would exist in them.) This conclusion will be picked up immediately below and go on to play a significant role in the development of the main argument of this chapter.

There remains the third question, how the existence in space and time of both initiators and appearances depends on our intuitions of the initiators. The answer is that initiators and appearances exist in space and time in logically distinct ways. First, although the existence of initiators *in space and time* does indeed depend on our intuitions of the initiators, the existence of initiators *simpliciter* does *not* depend on our intuitions of them.[2] We can separate their existence *simpliciter* from their existence in space and time on two grounds. First, the criterion of existence *in general*, as given in chapter 3, which is the notion of existence that is being applied to initiators *simpliciter*, is independent of any *particular* kind of intuition, including our own, and is therefore independent of our *a priori* modes of intuition.[3] Since these modes are space and time, the existence of initiators *simpliciter* can be separated from their existence in space and time. The second ground for the distinction is that, since space and time are merely our *a priori* modes of intuition, and since determinations that we intuit *a priori* cannot belong to initiators, and since space and time are such determinations, it follows that space and time cannot belong to these objects. Since just above we have also found that the existence of initiators in space and time is equivalent to the existence of appearances in space and time, and given that space and time cannot belong to initiators, if space and time are intuited *a priori*, it follows that space and time

2 This is not to gainsay the *criterion* of existence in general that was offered in chapter 3, namely, that an *existing* object can be given as an object through intuition. As I stated in the same chapter, section 1 fn., a *criterion* for the use or application of a concept is not meant to be the same thing as a sense or meaning of the concept.

3 To reiterate my interpretation of Kant's use of the concept of existence and its relation to the criterion of existence offered in chapter 3, as found specifically in section 6, since things in themselves cannot be intuited, at least not if they are taken as noumena in the *negative* sense of the term, the criterion of existence cannot be satisfied with respect to the application of the concept of existence to them. But this does not keep them from being merely *thought* as existing. And that is just what they are in Kant's theory of the possibility of *a priori* knowledge. It is a theory that states that knowledge requires intuition without itself requiring intuition, and hence without itself fulfilling the conditions of the possibility of knowledge. Hence, the theory allows for objects of thought alone, although, of course, it also allows for object that can be intuited, such as initiators, and object of intuition, such as appearances.

cannot belong to appearances either. So, if space and time are intuited *a priori*, they cannot belong to either initiators or appearances.

This conclusion reinforces the answer we gave to the second question raised above, namely, given the distinction between initiators and appearances, how can they *both* exist in space and time? The strengthened answer is that the existence of initiators *simpliciter* can be separated from their existence in space and time. On the other hand, we cannot similarly separate the existence of appearances from *their* existence in space and time, since space and time necessarily belong to them – a conclusion we also confirmed in the previous chapter, but originally arrived at in chapter 2. Appearances do not similarly exist *simpliciter*; they exist only in human intuition, and thus exist only as determined by space and time. So, while both initiators and appearances exist in space and time, existence in space and time is necessary only for the existence of appearances, but not for the existence of initiators.

[2] Dependence on Intuition and Presupposition. The second way in which initiators and appearances differ in their dependence on our intuitions of the initiators – the way that is pivotal for the remainder of our discussion in this chapter – is that the dependence of the existence of initiators in space and time on our intuitions of the initiators entails that if we do *not* intuit the initiators, they do not exist in space and time; but they would still exist, since their existence is independent of space and time. So, if we did not intuit them, it would be *simply false* that they exist in space and time. The existence of appearances in space and time, on the other hand, entails that if we do not intuit initiators, not only do appearances not exist in space and time, but they do not exist altogether. So, if we do not intuit initiators, it would *not* be *simply false* that appearances exist in space and time: To say, 'simply false' would be misleading, since to say that of a proposition usually implies, in a special sense of 'implies' (to be given immediately below), that the logical subject(s) of the proposition exists (exist), and we are presently considering a case in which logical subjects – appearances – do *not* exist. The nonexistence of the logical subjects of a proposition usually results in a different truth valuation from that of 'simply false,' which is ordinarily reserved for a proposition whose logical subjects *do* exist, but whose *predicate(s)* is (are) false of the subjects.

Recent analytic philosophy of language is largely responsible for the acceptance of the idea that the difference in the assignment of truth valuations such as just described in Kant's case can be the basis for drawing a distinction between two kinds of logical relation. It has been argued, notably by Frege and Strawson, and, I will shortly argue, by Russell as well, only more inconspicuously, that the non-existence of a logical subject of a proposition invites the valuation that the proposition is *neither true nor false.* All three philosophers call the logical relation based on this latter valuation *presupposition*, to distinguish it from logical implication or entailment, according to which, as a form of ordinary implication in general, whether logical or material, the falsehood of the consequent of the implication results in the *simple falsehood* of the antecedent, but not in the antecedent's being *neither true nor false.*

Accordingly, we will first explore the distinction between the two logical relations as it was developed by Frege, Strawson, and Russell. Since we have just seen how presupposition can play a crucial role in an interpretation of transcendental idealism (i. e. if we do *not* intuit initiators, propositions about appearances are *neither true nor false*), we can measure its use I am attributing to Kant against its uses by Frege, Strawson, and Russell.

[3] Presupposition and Analytic Philosophy. Kant is generally credited by analytic philosophers (which is a more general term than the aforementioned recent analytic philosophy of language) with providing a useful analysis of a relation between what is immediately given to us by means of the senses and what is thought in judgment. On the one hand, the senses gives us the material that is required for the application of our concepts to objects and our concepts provide the general rules for turning these objects into determinate objects of judgment. As Kant puts it in an aphorism that is especially dear to analytic philosophers: "Thoughts without content are empty, intuitions without concepts are blind" (A51/B75).

One reason analytic philosophers like Kant's analysis so much is that it parallels their own analyses of a relation between the senses and judgment. This relation has been a central topic of analytic philosophy since the time of early Russell. Although Frege considered the relation when he dealt with the corresponding relation between demonstrative or indexical expressions, or what Russell called "egocentric particulars," and general terms, he did so only in his later article, "The Thought,"[4] and at that

4 Frege, G., "The Thought: A Logical Inquiry," translated by A.M. and Marcelle Quinton, *Mind*, Vol. 65 (1956), pp. 289–311, reprinted in *Philosophical Logic*, ed. P. F. Strawson (London: Oxford University Press, 1967), pp. 17–38, esp. p. 24.

his treatment of the topic has to some seemed less than satisfactory.[5] Rather, it was Russell who actually introduced the topic into analytic philosophy, and he did so as an integral part of his theory of meaning.[6]

Russell laid down *sense-data* as the foundation of our knowledge of the external world. On the basis of what he called "logical analysis," he claimed that sense-data are the real logical subjects of our genuinely subject-predicate propositions – the propositions that express facts in which sense-data stand in various monadic and polyadic relations – expressed by the logical predicates of the propositions. Together, they make up basic, atomic level facts expressed by atomic propositions. By contrast, *descriptions* in *grammatical* subject positions ought to be analyzed instead as *predicates* of *second-level* existentially quantified propositions, which are built up from atomic propositions. The objects of these descriptions are just whatever objects satisfy these second-level predicates. Since the real initiators in the world are logical, not merely grammatical, subjects, objects of descriptions are not real, but are merely logical constructions out of what is real.

In a direct challenge to Russell's Theory of Descriptions, Strawson took grammatical subject descriptions to be just what they appear to be – subject terms. He thus urged us to reject Russell's analysis of them as disguised second-level predicates. Russell was right to look for a logical relation between grammatical subject-predicate propositions containing descriptions in subject position and quantified propositions to the effect that the described objects exist, but he did not get the right relation. He thought the relation to be that of *entailment* (which he called "logical implication"), whereas in actual usage it is *presupposition.* The difference between them, as already noted, is that the falsehood of an entailed proposition results only in the falsehood of the original proposition, whereas the falsehood of a presupposed proposition results in the original proposition's being neither true nor false. Frege, however, in "On Sense and Reference," was actually the first modern philosopher who recognized

5 See Kaplan, David, "Demonstratives," in *Themes From Kaplan,* (New York: Oxford University Press: 1989): 501, fn.26, and Perry, John, "Frege on Demonstratives, *Philosophical Review,* Vol. 86 (1977): 474–97.

6 Russell, B., "On Denoting," *Mind,* Vol. 14 (1905), reprinted in *Logic and Knowledge,* ed. R. C. Marsh (London: Allen & Unwin, 1956), *The Problems of Philosophy* (Oxford: Oxford University Press, 1912; reset 1957), *Our Knowledge of the External World* (delivered in Boston as Lowell Lectures, 1914, reprinted, New York, Mentor Books, 1960), and "The Philosophy of Logical Atomism," *The* Monist, (1918), reprinted in *Logic and Knowledge.*

this difference between the two logical relations and he used it in his analysis of grammatical subject-predicate propositions containing what he considered "names," which included descriptions.[7] Propositions containing Fregean names as grammatical subjects do not *entail* that the objects named exist; rather, they *presuppose* it. Consequently, if the named objects do not exist, the original propositions are neither true nor false, instead of being simply false. Unfortunately, Russell missed the distinction altogether, as Strawson tells the story, and ended up with the one relation he was working with – entailment. It is entailment, because for Russell if on logical *analysis* we find that we are really asserting an existentially *quantified* proposition when we ordinarily assert a *grammatical subject-predicate* proposition, the logical relation between the proposition we are analyzing – the analysandum – and the quantified existential proposition we find beneath the grammar – the analysans – is entailment, since the analysans is supposed to *logically follow from* the analysandum – which is one way of explicating entailment.

In fairness to Russell, however, it is important to note that in his "Lectures on Logical Atomism" he actually made the same distinction between entailment and presupposition that Frege had made and that Strawson was to employ in his attack on Russell. Russell even used the same name Strawson[8] was later to use for the relation – 'presupposition.'[9] Even Russell's reasoning for the need for the distinction followed Frege's thinking, not that his thinking wasn't independent of Frege's, but that he was not only aware of Frege's "On Sense and Reference," and had published in opposition to it, in "On Denoting." Indeed, the need grows out of a quite general problem.

We can assert the existence of an object corresponding to a *predicate* term of a subject-predicate proposition by means of an existentially quan-

---

7 Frege, G., "On Sense and Reference," translated by Max Black, in *Translations from the Philosophical Writings of Gottlob Frege*, ed. Peter Geach and Max Black (Oxford: Basil Blackwell, 1952), esp. pp. 63, 69. For the record, Strawson states that he arrived at his employment of the logical relation of presupposition and related issues in philosophy of language "completely ignorant" of Frege's work on the same relation and the same issues, despite the fact that his name is often paired with Frege's in this connection. See his, "Intellectual Autobiography," in *The Philosophy of P. F. Strawson*, ed. Lewis Edwin Hahn (Chicago and La Salle: Open Court Press, 1998), p. 7.

8 Strawson, P.F., "On Referring," *Mind* (1950), reprinted in *Essays in Conceptual Analysis*, ed. Antony Flew (London: Macmillan & Co. Ltd., 1956)

9 Russell, B., "The Philosophy of Logical Atomism," p. 204.

tified proposition in which the term again occurs as a predicate, and we can do so without confounding in any way the *reference* of the original subject-predicate proposition to an object. Thus, 'Something is yellow' can be used to assert the existence of an object corresponding to the word 'yellow' as it occurs in, say, 'Gold is yellow,' without affecting the reference of the latter proposition to gold. And something similar can be said about asserting the non-existence of such an object by means of a negative, existentially quantified proposition. Existential generalization with respect to predicates does not affect any independent reference to objects by means of subject terms.

The same cannot be said with regard to *subject* terms. We cannot use terms in subject position to assert the existence or the non-existence of objects corresponding to the terms without affecting the reference of the subject terms. For any such assertion of existence would be redundant, if not vacuous, and any such assertion of non-existence would be self-contradictory, since the object would have to be taken to exist if one were to assert its non-existence by using the term in question as the subject of the assertion. On this account, if 'gold' is the subject of both 'Gold exists' and of 'Gold does not exist,' the proposition 'Gold exists' would be vacuous while 'Gold does not exist' would be self-contradictory.

The general problem of asserting the existence or the non-existence of objects corresponding to *subject* terms was resolved in different ways by Frege and Strawson, on the one hand, and Russell, on the other. Though both sides employed the distinction between entailment and presupposition in resolving the problem, they did so on the basis of quite different theories of meaning. Frege's distinction between sense (*Sinn*) and referent (*Bedeutung,* alternatively translated as 'meaning') allowed him and Strawson to maintain that if an object corresponding to a subject term does not exist, the proposition containing the term can still have sense, even though it does not have a truth value, because the object is only the referent of the term, not its sense. For example, even though 'gold' is a name of a substance, it could still be a significant expression, even if gold did not exist. In that case, the proposition 'Gold is yellow' would be significant even though gold did not exist and the proposition would have no truth value.

For the sake of dealing with just a single example, let us assume, contrary to Russell's actual views, that 'gold' is a logically proper name as it occurs in 'Gold is yellow.' Russell would then hold, like Frege and Strawson, that the proposition presupposes the existence of gold. But then the divergence from Frege and Strawson begins. For if we tried to use 'gold' as a logically proper name and gold did not exist, since the existence of gold would be presupposed by the use of the name, the meaninglessness of 'Gold is yellow' that would result from the nonexistence of gold would be the end of the matter: There would be no sense (*Sinn*) attaching to 'gold' that would be left over once the term lost its meaning. Objects that are named constitute whatever content the names give to the propositions that contain them, since with respect to logically proper names, Russell rejects Frege's distinction between object and content, referent and sense. So, if gold does not exist and it is allowed to be used as a logically proper name, then 'Gold is yellow' would be missing some of what Russell would call its *meaning*, namely, gold itself, and thus the proposition would not be meaning*ful.* On that score, Russell would conclude, yet on quite different grounds from Frege and Strawson, that the proposition would be neither true nor false, which would accord with his adopting the general distinction between entailment and presupposition that he shared with Frege and Strawson. But, as can be seen, his semantic grounds for doing so would be radically different from theirs.

The above has been presented as a dichotomy between Frege and Strawson, on the one hand, and Russell, on the other. Frege and Strawson are said to have a view of sense (Frege's *Sinn*), that in determining reference (Frege's *Bedeutung*), it is distinct from it. Russell, on the other hand, repudiates the distinction. Thus the distinction appears central in marking the difference between the two semantics.

Evans and McDowell, however, dispute this view of the two semantics – a view that they recognize is held by many commentators on Frege, including Dummett and Burge.[10] They contend that though Frege (and I would add Strawson) explicitly adopts the position that names or descriptions of objects can have sense in the absence of the existence of the purported objects of reference (and therefore a thought with that sense as part of its content would not have a truth value), and thus would determine the objects of reference were the latter to exist, the position is not central to Frege's deeper views on the matter. In fact, it tends to mislead the interpreter of Frege, for it suggests, if it does not actually imply, that the determination is only one way – that the sense determines the refer-

10 Evans disputes Michael Dummett in Dummett's *Frege: Philosophy of Language* (London: Duckworth, 1973), ch 12, cited by Evans in *op. cit.*, pp. 12–13. And McDowell disputes Tyler Burge in Burge's "Belief *De Re*," in *Journal of Philosophy* 77 (1974), cited by McDowell in "*De Re* Senses," *loc. cit.*, pp. 215 ff.

ence, but not conversely. On the contrary, they argue, it is much more central to Frege's main semantical views that the object referred to (*Bedeutung*) and the context in which it occurs have a reciprocal determination on the sense (*Sinn*) in which the object, in its context, is presented. And this view is close to the one that is held by Russell on the matter.

Evans and McDowell are quite persuasive in their arguments against what they acknowledge has become standard interpretation of Frege's semantics. Since interpretation of Frege is not our present object, no attempt will be made to adjudicate the issue here. Given the controversy, however, two things must be said about the use of Frege in the interpretation of Kant that follows. First, as the Preface has already made clear, and as will become even more apparent in the next chapter, the Evans and McDowell interpretation of Frege not merely fits very well, but it actually supports, our use of Frege via our use of David Kaplan (in his "Quantifying In") in presenting our interpretation of Kant's uses of *de re* necessity. So, not only are Evans and McDowell not rejected in their interpretation of Frege, but are actually joined in our own interpretation of a Fregean style understanding of Kant. This will be especially evident in chapters 6 and 7.

But second, and as a point of departure from Evans and McDowell, there is a definite tilt toward Dummett and Burge in the interpretation of Kant's semantics with regard to the issue of presupposition and the withholding of truth value assignment in the case of the non-existence of putative objects of reference in Kant's propositions involving *de re* necessity. This tilt divides Kant between Fregean and Russellian semantics in sections 5 and 6 below. To sum up, not only is there no attempt to settle this controversy over the correct interpretation of Frege in the present circumstances, but each side to it is endorsed according to its respective, apparent closer proximity to Kant's own views on the matter as they are determined by the circumstances surrounding the discussion of Kant.

[4] Presupposition and Transcendental Idealism. As indicated in section 1 above, we can apply our discussion about entailment and presupposition to Kant's distinction between appearances and initiators. The proposition that initiators exist in space and time obviously presupposes the existence of initiators and entails the existence of objects in space and time. We have already seen in section 1 that the proposition *entails* that we intuit initiators. Hence, if we do not intuit initiators, nothing is in space and time, which makes the proposition that initiators exist in space and time simply false, but it leaves the presupposition that initiators exist unaffected. Again, in case we did not intuit initiators, the presuppo-

sition of their existence could still be true even though the implication that objects exist in space and time would be false.

The situation is more complicated, however, in regard to appearances. The proposition that appearances exist in space and time carries a presupposition similar to that carried by the proposition that initiators exist in space and time, viz., the existence of appearances. And it has a similar entailment as the proposition about initiators, viz., that some object is spatial-temporal. But as we have seen from the analysis just concluded, the existence of appearances, in contrast to the existence of initiators, *presupposes* that we intuit initiators, which in turn presupposes the existence of initiators. Therefore, if we do *not* intuit initiators, not only would the proposition that *initiators* exist in space and time be *simply false*, but the proposition that *appearances* exist in space and time would be *neither true nor false.* In other words, appearances differ from initiators with respect to their respective logical dependence on our intuition of initiators. The existence of initiators *simpliciter* does *not* depend on it, but their existence *in space and time* does, and in that case the dependence is an instance of *entailment.* Appearances, on the other hand, *do* depend on our intuition of initiators, and hence, so does the proposition that they exist in space and time, but in both cases the dependence is an instance of *presupposition* and not entailment. This concludes our recapitulation of the distinction between entailment and presupposition as applied to Kant in section 1.

[5] Kant's Russellian Semantics. In our discussion of Frege, Strawson, and Russell in section 2, we found that the difference in their uses of presupposition was based on their respective semantics. Frege and Strawson adopted the distinction between sense and meaning, whereas Russell rejected it. Which semantics might Kant's use of presupposition be based on – Frege's (and Strawson's) or Russell's?

All reference of our knowledge to objects would fail for Kant unless our knowledge ultimately referred to appearances, and we have seen that this presupposes that we intuit initiators. If we do not intuit them, however, not only are there no objects that our knowledge ultimately refers to, but, as already noted,[11] there are no intuitions that, as Kant puts it, "take place," either, since that requires that the objects are given to us, which in turn requires that "the mind is affected in a certain way" by the objects. This implies that we intuit initiators, since we intuit the objects that in affecting us are given to us. The conclusion is that intuitions do not "take place" unless we intuit initiators. Consequently, if we do not intuit

11 Chapter 2, sections 9 and 10, above.

initiators, the failure of our knowledge to refer to appearances could *not* have the particular Fregean style sense it might be supposed to be given by our intuitions of initiators if we did intuit them.

Kant's semantics would thus look like Russell's, at least with respect to the absence of Fregean senses that belong specifically to our intuitions of initiators. For Russell, names without objects have no content, since their content consists in their objects. Similarly, for Kant, representations without objects contain no intuitions that "take place." But though there cannot be intuitions that "take place" but have no objects, might there not be for Kant representations that have content but no intuitions that "take place"? If the answer is "yes," then the intuitions that "take place" that we do have would also carry that same content, that is, if the content determined in some way the intuitions that "take place." This would cast a Fregean aura over Kant's semantics.

[6] Kant's Fregean Semantics. We thus must direct our attention to our *a priori* intuitions of space and of time – intuitions that do not "take place," as Kant would put it – and to concepts. It should be mentioned that there is no point looking to *sensations* apart from their belonging to our intuitions or concepts, for, as determined in section 2 of chapter 3, they would then refer for Kant only to one's subjective states and not to a state of an object (A320/B376), and we saw in the previous chapter that Kant does not consider such wholly subjective reference to consist of content. Sensations become content only by being contained in intuitions or concepts. Intuitions provide the spatial and/or temporal relations in which sensations can be "ordered and posited," and thus referred to objects, Kant claims (A20/B34).

Empirical concepts may correspond to empirical intuitions that "take place" or they may be imaginary. Those that correspond will have spatial and/or temporal features corresponding to features belonging to the intuitions. And those that are imaginary will *also* have spatial and/or temporal features, since without them the concepts would, like the sensations they contain, refer only to one's subjective states, and therefore, again like their sensations, have no content.

We must therefore direct our search for content to our *a priori* intuitions of space and of time and to *a priori* concepts. We can then dispense with the *a priori* concepts of the understanding, i. e. the categories, on the ground that presently we are concerned with Kant's use of presupposition, not entailment. To be specific, the categories carry no presupposition of the existence of objects so thought, since for Kant they are concepts of objects *in general*, which we have seen carry no such presupposition. Moreover, if they did carry such a presupposition, it would undermine the very legitimacy, indeed, the very need, of their Transcendental Deduction. For the Deduction is supposed to demonstrate that and how concepts that determine how we can think objects in general have instances in the objects given to us through the independent faculty of our sensible intuition. Therefore, since we are searching for possible Fregean style content in Kant's specific semantics of presupposition, we are left with the single question of whether our *a priori* intuitions or concepts of space and time have content.

We can briefly dispense with our *concepts* of space and time since they depend for Kant on our *intuitions* of space and time and can only be understood in terms of those intuitions. Therefore, we can turn to our *a priori* intuitions of space and time to see if they have content.

We find, however, that if they did *not* have content, it would be impossible for them to refer to empirical intuitions, and thus refer to appearances. Without the possibility of a reference to appearances, appearances would constitute whatever content our *a priori* intuitions might have. But that would keep the intuitions from being *a priori*, which would allow Kant's epitomizing question of how synthetic *a priori* judgments are possible to be abruptly dismissed on the grounds that *a priori* intuitions cannot refer to appearances. Since the possibility of synthetic *a priori* judgments consists in their reference to appearances, which in turn requires the possibility of *a priori* intuitions, if the intuitions for their part cannot refer to appearances, the judgments that depend on them for *their* reference to appearances would also be impossible. In a word, since judgments refer to appearances through intuitions, both *a priori* and empirical, if *a priori* intuitions could not refer to appearances, neither could the judgments, and that would be the end of the matter.

[7] Difficulties of a Perfect Fit. Although we may conclude that Kant's *a priori* intuitions must have content, in the sense of a Fregean style sense that can be distinguished from a reference or meaning, these intuitions so far lack the other property belonging to Frege's names, viz., the presupposition of the existence of the reference that the name purportedly refers to. In this regard, Kant's *a priori* intuitions are no different from his categories, since *a priori* representations in general carry no presupposition of the existence of the reference of such representations. That reference consists of appearances, and no *a priori* representation can by itself be logically related to appearances, whether the logical

relation is presupposition or entailment. Reference to existing objects for Kant is ineliminably empirical.[12] So, while we have found content in Kant's *a priori* intuitions, it is so far not the sort of content Frege and Strawson attributed to referring expressions, viz., content that describes objects whose existence is *presupposed* by the use of the expressions.

This shortcoming can be easily remedied, however, in two, related steps. Earlier we noted that if we could find that our *a priori* intuitions have content, and agree that for Kant the intuitions *determine* empirical intuitions, and now we can add, determine them with respect to what Kant calls their form, then the empirical intuitions would have the content of the *a priori* intuitions. So determined, empirical intuitions would be spatial/temporal representations. Secondly, since empirical intuitions alone presuppose the existence of appearances, *a priori* intuitions, while not themselves presupposing such existence, can be said to be part of such a presupposition, if it can be proved that they *refer* to empirical intuitions, and through them *refer* to appearances. Kant calls such proofs transcendental expositions of the *a priori* concepts (the nomenclature of "exposition" [*Erörterung*] applies only to the *concepts* of space and time instead of to the *intuitions* of space and time). So, the first step consists in demonstrating that empirical intuitions must be *determined by a priori* intuitions, and that determination can be found in the spatial/temporal *forms* of the empirical intuitions. The second step consists in proofs that the *a priori* intuitions *refer* to empirical intuitions and thereby refer to the appearances whose existence is *presupposed* by the empirical intuitions. In sum, *a priori* intuitions must both determine and refer to empirical intuitions. *A priori* intuitions can then be said to be part of the *presuppositions* belonging to empirical intuitions.

There is only one hitch in this attempt to make Kant's semantics fit Frege's. The hitch goes directly back to our discussion of Russell. We there found that it was impossible for empirical intuitions to have a Fregean sense in case of a failure of the presupposition of the existence of the appearances that correspond to the intuitions. This would happen, we found, in case we did not intuit *initiators* whose appearances were presupposed by the intuitions. We saw that a gap would be left by our failure to intuit those initiators. No Fregean sense would remain in that instance,

---

12 Strawson would say "demonstrative" or "contextual" and Russell would say "indexical" or "egocentric." See Strawson, *Individuals: An Essay in Descriptive Metaphysics* (London: Methuen & Co. Ltd., 1959) ch. 1, section 1., and "On Referring," and Russell, *The Philosophy of Logical Atomism.*

since the loss of the appearances would result in the loss of the empirical intuitions themselves – the failure of the intuitions to "take place," as just noted – and therefore would result in the loss of any Fregean sense that would otherwise attach to the intuitions. Whereas we can compare the missing empirical intuitions to Russell's spurious logically proper names, that is, names without meaning, there is nothing comparable in Frege's semantics, since Frege's and Strawson's referring terms still have sense despite the non-existence of any presupposed reference. As we remarked earlier, the only senses left over for Kant in cases in which we do not intuit initiators are the senses of the *a priori* intuitions that would otherwise determine and refer to the empirical intuitions, that is, if the empirical intuitions had "taken place."

[8] Kant's Mixed Semantics. So, we are at an impasse in our attempt to make Kant's semantics fit either Frege's (and Strawson's) or Russell's semantics: As soon as such a fit looks promising, it is dislodged by the intrusion of the other semantics. If empirical intuitions presuppose the existence of appearances, à la Russell, they cannot be said to have any Fregean sense that can be distinguished from the appearances; or, if they have sense that remains present in case the appearances do not exist, it belongs only to the *a priori* intuitions that would otherwise have determined and referred to the empirical intuitions, had the latter "taken place." But we have seen that *a priori* intuitions of themselves do not presuppose the existence of appearances; they must determine and refer to empirical intuitions to do so. Consequently, the dilemma for empirical intuitions is that either they have Fregean sense (carried by *a priori* intuitions) but do not presuppose the existence of appearances or, à la Russell, they presuppose the existence of appearances but have no Fregean sense. In sum, Kant's semantics is not a perfect fit for either Frege's or Russell's semantics.

[9] Russell's "substrates" and Kant's "initiators." Before leaving our oscillating image of Kant as now Russellian, now Fregean, we can introduce a further item that will conclude our discussion of Kant's relation to the semantics of both Frege and Russell. We have noted that the spatial/temporal properties that Russell attributes to sense-data are considered by him to be predicative, not intuitive, real relations in which the sense-data stand as relata. Other properties of sense-data, such as colors, touches, and sounds are also understood as relations – monadic relations – in which the sense-data stand as particulars. Once even sensations are understood as relations, however, it seems that the sense-data themselves have no properties of their own, which would make them "bare particulars." It

was only after he gave his "Lectures on Logical Atomism" that Russell took exception to this position. In his *Inquiry into Meaning and Truth*, he acknowledged that the earlier position had unfortunately committed him to the notion of the sense-datum as "an unknowable subject," or metaphysical *substrate.*[13]

His later philosophy disavowed the substrate and instead based our attribution of properties to particulars on a lower level of language that is free of particulars altogether.[14] Such languages have been called property-location languages. They tend to turn sortal terms into mass terms, thereby ridding language of terms for individuating particulars, and instead allow only terms that can express the presence, absence, increase, or decrease of a property at a location. For instance, one might find such expressions as "woman here, now," "more woman" or "less woman" instead of "same woman" or "another woman." Quine's *occasion sentences* and his purportedly empirically objective *observation sentences* in particular, which are crucial to his theory of language in *Word and Object* [15] as well as in other works of his, are property-location languages, as are Strawson's *feature-placing sentences* which figure prominently in the second part of *Individuals.*[16]

Kant would have demurred from Russell's disavowal of substrates in favor of a property-location language. As bad a name as substrate has received in empiricist philosophy, especially since Locke, Kant would still have remained committed to an inexact counterpart to it in his notion of a *thing*, which in this book we have interpreted as *initiator* (but left as *thing* in *Kant's Theory*) and have already seen to play such a strategic role in his theory of knowledge. The counterpart to substrate is only inexact because, first, in one respect, an initiator is not unknowable at all; it is unknowable only with respect to properties and relations that it has independently of sensibility; but it is quite knowable with respect properties and relations belonging to sensibility, i. e. spatial-temporal properties and relations and qualities that can be ordered in space and time. Second, an initiator is not without properties or relations of its own, i. e. it is not a bare particular, as is a substrate; rather, it is just that we cannot know

13 Russell, B., *An Inquiry into Meaning and Truth* (New York: W.W. Norton, 1940), p. 98.

14 Ibid.

15 Quine, W. V., *Word and Object* (New York: The Technology Press of the Massachusetts Institute of Technology and John Wiley & Sons, Inc.: 1960), Chapter 11.

16 Strawson, P. F., *Individuals,* Chapter 6.

these intrinsic properties and relations (since they constitute things in themselves). Yet initiators would still stand as Kant's counterparts to substrates. For properties and relations belonging to sensibility, which are attributable only to appearances of initiators, and those properties and relations intrinsic to initiators, which are attributable only to initiators (things) in themselves, are mutually exclusive sets of determinations of initiators; therefore, the identical objects – the initiators – that can be determined by both sets of determinations must be thought independently of both of them, as we do, viz., as external objects that get our knowledge started, which we go on to identify as objects that in affecting us, are given to us. That is, the term 'initiator' has been employed independently of any reference to the two sets of determinations just mentioned – those belonging to sensibility and those that are intrinsic to the objects. Moreover, the identification of initiators as objects that in affecting us are given to us is also independent of these two sets of determinations. Therefore, in the sense that they support both sets determinations and yet are thought independently of them, as that independence has just been explained, initiators would be counterparts to substrates, albeit inexact counterparts.

[10] Summary. Our investigation of the relation of Kant to analytic philosophy has taken us from one of analytic philosophy's favorite dictums of Kant's, that "thoughts without content are empty, intuitions without concepts are blind," to Russell's epistemological and logical foundations of our knowledge of the external world, to Frege's and Strawson's alternative theories that entail a sense-reference distinction and then back to Russell's own use of the distinction between presupposition and entailment that eschewed any sense-reference distinction. From there we went on to Kantian exegesis and to the application of the distinction between entailment and presupposition to that exegesis.

Part II: Necessity

Chapter 6 – Kant's Referential Ambiguity

[1] Kant's Problem of Quantifying In. In chapter 2 we found that, for Kant, initiators are independent from appearances, but not conversely. Consequently, in Kant's account, there must be a condition under which initiators are so referred to, or given to us, that they can be represented as appearances. This condition is sensibility; hence, initiators must be referred to by us by means of sensibility if they are to be represented as appearances. Since initiators can be represented as appearances only if we (empirically) intuit them, the same condition of sensibility also determines our intuitions of them. That is, for Kant, we can intuit initiators only by means of sensibility, if they are to be represented by us as appearances. In other words, the intuitions through which we intuit initiators are the references we can make to them by means of sensibility. So, in Kant's theory of our *a priori* knowledge of objects, we have objects (initiators), the means by which we can refer to them intuitively (sensibility), representations through which we can refer to them (intuitions), and representations through which they can be suitably represented as of initiators (appearances). To summarize, it is through sensibly intuitive references to initiators that they can be represented as appearances.

Moreover, since initiators and space and time are independent of each other, to the extent that appearances depend on initiators, there must be a condition under which both initiators and appearances are so referred to that they can be represented as existing in space and time. This would therefore also be a condition on references to initiators and appearances under which space and time are represented as containing existing objects. In addition, we found that condition to be that we intuit initiators by means of outer and inner sense, respectively. In sum, initiators must be subject to sensibility to be represented through empirical intuitions as appearances and both they and appearances must be subject to outer and inner sense, respectively, to be represented as existing in space and time.

Similarly, since sensibility and outer and inner sense, and therefore space and time, are independent of the understanding, and conversely, the understanding is independent of them, and since appearances represented as existing in space and time, or simply, appearances in space and time, depend on sensibility and inner and outer sense, and the categories, on the understanding, appearances in space and time are independent of the categories, and conversely, the categories are independent of both space and time and appearances in space and time. Consequently, there must be a condition under which appearances in space and time are so referred to that they can be represented as categorized, or simply, appearances in space and time are so referred to that they can be categorized; this would therefore also be a condition under which the categories *extend to* (*auf . . . gehen*) objects of *a sensible intuition in general* (*überhaupt*).

The Transcendental Deduction of the categories argues that there are further conditions under which the categories are related to objects. First, there is the condition under which objects of a sensible intuition *in general* are not merely within the extent, or reach, of the categories, but are *determined* (*bestimmt* [B128, B143, B161, *et passim*]) by them, viz., the 'I think'. This is the *intellectual* condition under which objects are determined by the categories. And second, there is the condition under which objects that can be given to us through *our* sensible intuition in particular, i.e. appearances, can have the categories *applied to* (*auf . . . angewandt* [B161]) them. This is the imaginative condition – the imagination – under which *appearances*, and not just objects of a sensible intuition in general – are determined by the categories. Finally, the imagination in turn is divided into, first, the *a priori*, or *productive*, imagination (B151), and second, the *empirical* imagination that plays a role in the account of the possibility of *a priori* knowledge, viz., the synthesis of apprehension (B160). The latter is meant to exclude the so-called *reproductive* imagination, which belongs only to psychology, but which unfortunately for Kant played a prominent role in the first edition of the *Critique.* Since our interest is not in the Transcendental Deduction but only in the ascription of the categories to appearances as determining their necessary interrelations in space and time, and since the imaginative conditions are determined by the intellectual condition of the 'I think,' for the sake of a simplified exposition, all these conditions will henceforth be abbreviated by the single condition that determines all of them, viz., the intellectual condition, i.e. the 'I think,' even though this condition is not *sufficient* to determine the interrelations among appearances in space and time – for that we also need the imagination, not to mention, ultimately, a representation that is already given, from sensibility alone, i.e. empirical intuition.

To sum up, there must be a condition (i.e. sensibility) under which initiators are referred to, or given to us, such that they can be represented as appearances; references to both initiators and appearances must be under a such a condition (i. e. outer and inner sense) that they can be represented as existing in space and time; and there must be a condition (a sensible intuition *in general*, and hence the 'I think') on references to appearances in space and time under which they can be represented as categorized.

But such conditions on our references to objects must be strong enough for Kant to conclude, not just that initiators can be or are represented as appearances, but that they *must* be represented as appearances; conditions on our references to appearances must be strong enough to conclude, not just that appearances can be or are represented as existing in space and time, but that they *must* be so represented. Similarly, the conditions on our references to appearances in space and time must be strong enough to conclude, not just that they can be or are represented as categorized, but that so referred to, they *must* be represented as categorized. Otherwise, Kant would fail in his attempt demonstrate *a priori* that our *a priori* knowledge refers to appearances, that is, demonstrate that our *a priori* knowledge *must* refer to appearances. Kant would have no transcendental expositions or deduction of concepts, whether the concepts are those of space and time or the categories, respectively. In other words, Kant would fail in his main positive objective in his *transcendental* epistemology. The conditions of sensibility, outer and inner sense, and being subject to a sensible intuition *in general* are supposed to be strong enough to yield the necessities in question. Kant maintains that his argument in support of his claim that *they are the a priori means by which we represent objects* is sufficient to assure these necessities. This claim, he contends, is itself *a priori*, although he maintains that there is no *a priori explanation* of why we refer to objects through these particular *modes of reference*, i. e. space, time, and the categories (that are connected to our *means of representation*, i. e. sensibility, outer and inner sense, and a sensible intuition *in general*, and hence the 'I think,' and not other modes of reference (that presumably might be connected to these or perhaps other means of representation)) (B145–46).

[2] *De Re* and *De Dicto* Necessity vis a vis Kant. It may not need repeating that the necessity referred to in the immediately preceding paragraph – indeed, the necessity with which we are concerned through the book – is *de re* necessity. It is the necessity of properties or relations that are ascribed to an *object* or to *objects:* They are necessary properties and relations of objects. It stands in contrast to *de dicto* necessity, where the necessity is ascribed to a *proposition – to what is being said*, in contrast to saying of *objects* that they have necessary properties or relations.

In respect to the question of whether Kant employs *de re* or *de dicto* necessity in formulating his transcendental idealism, the present work is in agreement with Paul Guyer, that the necessity is *de re:* e.g. with respect to appearances, the necessary intuitions of space and time belong to them, and with respect to spatial-temporal appearances, the necessary concepts of the understanding – the categories – necessarily determine them (e.g. A38/B55).[1] And, it should be added independently of Guyer's position that, if we intuit initiators, they must be represented, or given to us, as appearances. The *de re* character of the necessity reflects the independence of the reference to the objects from the necessary properties or relations that are being ascribed to them. Initiators are referred to independently of the necessity with which they are represented, or given to us, as appearances; moreover, as objects of our empirical intuitions appearances are referred to independently of any ascription of necessary spatial and temporal relations to them; and, finally, the Transcendental Deduction is premised on the possibility that appearances in space and time might *not* be categorized, and that possibility must be bypassed in some way. None of these necessities is *de dicto*, and none is analytic. In every case, references to the objects, whether to initiators, appearances, or appearances in space and time, stand *outside*, or are 'external', to the necessary properties or relations that are ascribed to the objects,[2] where the term 'external' is used in a sense that is logically related to the sense that was given to it in sections 4 and 5 of chapter 2.

The problem of possible opacity for Kant is created by this very logical independence of the references to the objects from the necessary properties and relations that are being ascribed in the contexts of *de re* necessity. How can the properties or relations be necessary with respect to the objects if the objects are being referred to independently of their predication? Kant's answer in each case is that the ascription must be *conditional.* Initiators must be given to us as appearances, *if we intuit them by means of sensibility*; appearances must exist in space and time, *if they are objects of our outer and inner sense,* respectively, as well as objects of our empirical intuitions; and appearances in space and time must be catego-

1 Paul Guyer, *opus. cit.*, p. 363 ff. and pp. 379—80.

2 Thus, the operator has what is called "narrower scope." For a discussion of the distinction between "narrower" and "wider" scope of a modality such as necessity, see, for example, Arthur F. Smullyan, "Modality and Description," *The Journal of Symbolic Logic*, 1948, Vol. 13, reprinted in L. Linsky, ed., *op. cit.*

rized, *if they are objects of a sensible intuition in general, and hence are subject to the 'I think'*.

Yet, two exceptions to Guyer's work must be made. When discussing *de re* necessity, Guyer claims that Kant's use of the modality is absolute, and not conditional. Yet the use of *de re* necessity attributed to Kant here is explicitly conditional, as the last paragraph, among others, attests. Necessities are no less *de re* for being *conditional* necessities – something that Guyer does not seem to countenance. So, on that score, exception must be taken to his work.

The second exception, and it is really only a merely possible exception, since Guyer does not actually address the issue, is that he does not distinguish between two types of *de re* necessity as others do. Kit Fine points out, for example, that Quine's skepticism about *de re* necessity can be divided between a *logical* or *syntactic* issue involving the principles of substitution and existential generalization in modal contexts and a *metaphysical* issue involving the possibility of making intelligible the idea of the necessary fulfillment of a condition, i.e. the issue of traditional Aristotelian essentialism.[3] *The interpretation of Kant's use of de re necessity offered in this book pertains only to the first type of the necessity.* Of course, when Kant employs that particular necessity to justify the use of the second type, the type associated with Aristotelian essentialism, whatever arguments are presented here regarding his use of the first type of necessity must apply to his use of it when he tries to justify our use of the *second* type. The latter can be found in both the Transcendental Deduction, where he justifies our use of the category of *substance* as part of his justification of our use of all of the categories, and the First Analogy of Experience, which states a principle determining the use of the concept of *substance* in experience.

[3] Quantified Objects Cannot be the Objects Conditioned References Refer To. The preceding section first refers to independent, or external, objects, first, as initiators (as logically independent of appearances), then as appearances (as logically independent of space and time), and finally, as appearances in space and time (as logically independent of the

3 Kit Fine, "The Problem of De Re Modality," in *Themes from Kaplan,* ed. J. Almog, J. Perry, and H. Wettstein (New York: Oxford University Press, 1989), pp. 197–202, for a development of these two types of necessity. David Kaplan makes the same point as Fine does about the distinction between logical and metaphysical necessity and develops the point in his essay, "Opacity," in *The Philosophy of W. V. Quine,* ed. by Lewis Edwin Hahn and Paul Arthur Schilpp (La Salle, IL: Open Court, 1986), pp. 249 ff.

categories). It then speaks of them as being represented as having properties or relations. Accordingly, initiators are represented as appearances, appearances as spatial and temporal, and spatial and temporal appearances as categorized. Penultimately, it conditions the references to the objects so that it can speak of them as having necessary properties or relations, – as necessarily appearances, as necessarily spatial and temporal objects, and as necessarily categorized. *Seriatim*, the conditions are sensibility, outer and inner sense, and the 'I think'. Finally, it speaks of the first items – the objects (initiators, appearances, and appearances in space and time) – as being logically independent of these conditions of our references to the objects. So, initiators are logically independent of sensibility, appearances, of outer and inner sense, and appearances in space and time, of a sensible intuition *in general,* and hence of the 'I think'.

Imposing these conditions on references, we dispel Quine's worry about opacity. Since initiators remain the same whether they are referred to by means of sensibility (as appearance) or not, and since, if they are referred to independently of sensibility, it is false that they *must* be represented as appearance, the use of the *de re* necessity just employed depends on the role of sensibility in our references to them, viz., that it conditions our references to them. By conditioning our references to initiators by means of sensibility, we avoid inducing opacity when we state that initiators can be represented as necessarily being appearances.

In the proposition that initiators can be represented as necessarily being appearances, or that initiators are necessarily appearances, it is not initiators *simpliciter* that we are referring to, if we are to avoid opacity, but rather *initiators as they are referred to by means of sensibility.* This is an instance of Quine's truth that the necessity in question depends on the mode of reference to the object. His point is that predications of necessary properties or relations of references to objects requires that the references be suitably conditioned, if opacity is to be avoided. Similar remarks are applicable to references to appearances and their representations as necessarily spatial and temporal and to references to spatial and temporal appearances and their representations as necessarily categorized. The former must be conditioned by outer and inner sense respectively, and the latter, by a sensible intuition in general, and hence by the 'I think'.

Although Quine is right that absent such conditions on the references, opacity is induced by the use of necessity, it is now widely recognized that he is mistaken in thereby concluding that the presence of such conditions entails traditional Aristotelian essentialism. The presence or absence of such conditions is independent of the question of traditional Aristotelian essentialism, and conversely. So, the issues of opacity, or quantifying into necessity contexts, and those of traditional Aristotelian essentialism are independent of one another. Consequently, Quine's concern with opacity remains an issue that must be resolved by our account of Kant's uses of *de re* necessity even though we can separate it from the question of his commitment to traditional Aristotelian essentialism. The issue for us, therefore, is finding *a perspicuous formulation of Kant's solution to the problem of opacity.* We have shown how he uses his theory of conditioned references to objects to explain how, *as objects whose references are thus conditioned,* they can be represented as having necessary properties and relations. Our question now is, can that ascription, or representation, of necessary properties and relations be made more perspicuous, if we give Kant's uses of *de re* necessity a completely *general* interpretation? The more general the interpretation, the more powerful will be the explanation of Kant's uses of *de re* necessity.

We know that if the objects that are referred to are represented as having necessary properties and relations without the references to them being suitably conditioned, we come up against opacity, since the references will be open to substitutions that falsify the use of necessity. Our general interpretation of Kant's uses of *de re* necessity must therefore first contain a place for a condition on references to objects. If the objects are initiators, the interpretation must have a place for sensibility as a condition on references to the initiators, and similarly, if the objects are appearances, the interpretation must have a place for outer and inner sense, and if the objects are appearances in space and time, the interpretation must have a place for a sensible intuition *in general,* and hence for the 'I think'.

Having created a place for conditioned references in our general interpretation of Kant's uses of *de re* necessity, we turn to the *first* of two major questions he faces in his uses of the necessity: What are the objects that enter into (i.e. that are quantified into) the necessity contexts? That is, how are we to interpret Kant's saying of certain objects that they have certain necessary properties and relations – interpret in such a way that we do not induce opacity? Let us call these objects the *quantified objects* of the interpretation.

Let us suppose that the quantified objects are, *seriatim,* the initiators, appearances, and appearances in space and time – henceforth called collectively *initiators, et al.* – that the conditioned references refer to. On this hypothesis, however, opacity *will* be induced, as Quine has warned. An indirect argument can make this clear. The hypothesis is that the quantified objects are initiators, *et al.* So, it is initiators, *et al.* that are quantified into the necessity contexts. Under the hypothesis, it would not matter which references are made to initiators, *et al.* for the modal propositions to be true, since it is initiators, *et al.* themselves that are quantified into the necessity contexts. Any references to the objects could be substituted for any other references to the same objects. This is why the principle of substitution is a test for reference to objects, and thus why violation of the principle results in opacity. And if any reference to the objects would do, no particular *mode* of reference would be necessary. But since we have agreed with Quine that references to objects must be *conditioned* by particular *modes* of reference, if they are to be involved in non-opaque, i. e. transparent, modal propositions, it follows that *the quantified objects cannot be the same objects the conditioned references refer to.* It further follows that, for Kant, initiators, *et al.* cannot be the quantified objects. Kant cannot quantify over initiators, *et al.*, if he is to avoid opacity in necessity contexts. That is part of the lesson of transcendental idealism: that we cannot have *a priori* knowledge of initiators unconditioned by sensibility, that is, initiators *in general* (*überhaupt*), and thus initiators in themselves (A35/B51–2).

The point can be made a bit differently. As long as references to initiators, *et al.* are unconditioned, we can refer to the initiators, *et al.* independently of any reference to their being referred to at all, since it would not matter which references might be involved in our references to them. Any reference would do. And if that were so, no *mode* of reference to the objects nor any *particular* reference to the objects would be necessary for us to refer to the objects. That is precisely what Quine is aiming for and what makes the objects themselves, initiators, *et al.*, the objects of our references. But then, if we tried to quantify into necessity contexts, we would induce opacity.

[4] Quantified Objects As Objects of Conditioned References. Since the objects the conditioned references refer to cannot be Kant's quantified objects, nothing else involved in the references might qualify as the quantified objects except the conditioned references themselves. Let us therefore examine the hypothesis that indeed the quantified objects in question are the conditioned references themselves.

On this hypothesis, since these conditioned references are sensible representations through which we refer to initiators in particular, the objects that enter into the construction that *x* is necessarily an appearance would be these representations. For example, the representations that we talked about in section 9 of chapter 2 and in sections 5 and 6 of the previous chapter, viz., intuitions that "take place," which cover empirical intuitions, would be among such conditioned references to initiators. In that instance, empirical intuitions would be the objects that enter into the contexts in which it is said of objects that they have the necessary property of being appearances. That means that in the interpretation, intuitions that "take place" would occur both *without* and *within* the necessity contexts.

Let us for the sake of simplicity restrict our discussion of the nature of the quantified objects for the time being to initiators and the predication of the necessary property of being an appearance. One might then object to this interpretation of the quantified objects of Kant's uses of *de re* necessity. The interpretation seems to have reversed Kant's actual thought about the matter. The hypothesis that the quantified objects are the conditioned references seems to be saying that what are necessarily appearances are the conditioned references themselves, which include empirical intuitions. Yet, appearances are initially defined as the (undetermined) objects of empirical intuitions. So, it is appearances that are the objects of empirical intuitions, not the other way around, as the hypothesis seems to indicate.

The objection is supported by the following consideration. Empirical intuitions refer to initiators, and if we take the representations of the objects that occur in our empirical intuitions of the initiators to be appearances, we can let appearances be the objects of empirical intuitions. As reiterated in section 1 above and first argued in chapter 2, appearances cannot be *identical* with initiators, since, *inter alia*, it would be vacuous to say that initiators occur in our empirical intuitions of the initiators. Therefore, we cannot let the representations of initiators that occur in our conditioned references to initiators (i.e. appearances) be the *quantified objects* that are referred to within the necessity contexts, if *ex hypothesi* the quantified objects are the conditioned references themselves, i.e. including empirical intuitions. Consequently, we cannot interpret Kant's quantified objects as the conditioned references to initiators, *et al.*

A third point supports the objection. Since the hypothesis seems to be saying that what are necessarily appearances are the conditioned references themselves, it must be wrong, since it is the *initiators-as-referred-to*, or as they occur in the conditioned references to them, or, again, as the occurrences of the initiators in the conditioned references to them, not the conditioned references themselves that refer to them, that are necessarily appearances, if the references to them are suitably conditioned, i.e. by means of sensibility.

The objection seems to be based on a misunderstanding, however. The objection seems to take the conditioned references as they occur *within* the necessity contexts (which is how they should be taken, since the quantified objects whose nature we are trying to determine are supposed to be identified with them) as items that are *mentioned,* but not *used,* within the contexts. Accordingly, the objection understands the hypothesis regarding the interpretation of the quantified objects as a proposition about the conditioned references themselves, and as predicating of *them, per impossible,* that *they* are necessarily appearances. On the contrary, however, within the contexts in which a necessary property or relation is being ascribed to an object, such as necessarily being an appearance, the conditioned references are *used.* So, while the quantified objects whose interpretation is in question are still identified as the conditioned references, what those conditioned references refer to *within* the necessity contexts and about which the proposition asserts that they are necessarily appearances are yet *further* objects.

[5] Further Objects and Kantian Ambiguity. This raises the question of which objects these further objects might be? This is the *second* of the two major questions that were alluded to above regarding Kant's uses of *de re* necessity (the *first* being, What are the objects that enter into [i.e. that are quantified into] his necessity contexts? and the answer was: the conditioned references themselves). We have just seen that these further objects cannot be the conditioned references themselves, since the conditioned references themselves refer to these further objects and it is being said of these further objects that they are necessarily appearances, whereas the conditioned references themselves are not appearances at all. Kant distinguishes appearances from their sensible representations, including empirical intuitions, as the objection correctly insists.

Might these further objects be the *original* objects that the conditioned references refer to outside the necessity contexts, i. e. the initiators? If so, we are back with the problem faced earlier, viz., that it would then not matter *which* references were made to the initiators, and this would vitiate any force that the idea of conditioning might have in Kant's theory that our references to initiators must, on pain of opacity, be conditioned by sensibility. So, the further objects can't be either the conditioned references themselves or what they originally refer to, i. e. initiators.

In the search for these further objects, we can now return to our more general discussion about conditioned references to initiators, *et al.*, and the ascription of necessary properties and relations to objects. To determine which objects these further objects might be, we must take a certain view of how Kant's idealism treats conditioned references in general. In a word, they occur in the idealism *ambiguously: Within* contexts of *de re* necessity, they do not refer to the same objects that they refer to *outside* the contexts. Outside the contexts, they refer to initiators, *et al.* Within the contexts, however, they refer to the initiators, *et al. as they are represented through the conditioned references to them that occur outside the contexts.* Accordingly, the objects of the conditioned references *within* the contexts are *representations* of the initiators, *et al.* relative to, in the sense of the references' referring back to, the conditioned references that are made to the initiators *outside* the contexts. Transcendental idealism thus contains *within* necessity contexts references to initiators, *et al.*, but only as they are referred to through the conditioned references as they occur *outside* the necessity contexts, and therefore, only as they are subsumed under the conditions on the references that are made to them.

Alternatively, expressed in a manner that is perhaps closer to Kant's way of expressing the idea, transcendental idealism contains *within* necessity contexts references to initiators, *et al.* only as they *stand in relation to* the conditions on the references that occur *outside* the necessity contexts. Since these are the same conditions that determine the references as they occur *within* the necessity contexts, it follows that, in explanation of the *ambiguity* in question, the references *within* the contexts refer to initiators, *et al. only as they stand in relation to* the conditions that determine the references. Outside the necessity contexts, initiators, *et al.* are *not* conditioned by the conditions on the references, but *within* the necessity contexts, they *are.* As conditioned, initiators, *et al.* are *representations* of themselves. These *representations* are offered, then, as the *further* objects we have been looking for. They are the objects that the quantified objects, i.e. the conditioned references, refer to *within* the necessity contexts.

To make the point in a summary way, a way that borrows from Kant's own language on the matter, the conditioned references *within* the necessity contexts refer to the objects only *as they are represented*, viz., the objects *as they are referred to*, through the same conditioned references that occur outside the necessity contexts. Outside the necessity contexts, the conditioned references refer instead to the objects themselves, *simpliciter*, i.e. initiators, *et al.*, even though the references are nonetheless conditioned. In the cases of references conditioned by sensibility, the references must be *sensible.* And in the case of references conditioned by the understanding, the references must be *intellectual*, and as we have also seen, *imaginative* as well.

This is why it was said above that the objects that occur within necessity contexts are not the same as those that occur outside such contexts. Outside the contexts, conditioned references refer to initiators, *et al.*, *simplicter*, though still referred to through conditioned references as well as independently of the conditions. But inside, they rather refer to initiators, *et al.* only *as they are referred to through the conditioned references*, or only as objects *of* the conditioned references. This explanation of the *further* objects in question is meant as an interpretation of Kant's idea that appearances belong to the real existing objects (i.e. initiators) the appearances represent, but only insofar as the real existing objects stand in relation to sensibility. Therefore, appearances belong to sensibility as well as to the objects. We can thus say that being an appearance is a *relational* property of initiators insofar as they stand in relation to sensibility. Only as thus relativized, says Kant, can necessary properties or relations be ascribed to the initiators. Consequently, the objects themselves, i.e. the initiators, are not the objects of the ascriptions; rather, they are the objects of the ascriptions only insofar as they possess this relational property of being an appearance. That, on the interpretation offered here, is the gist of Kant's theory that judgments about things (initiators) that are conditioned by sensibility refer to appearances rather than things (initiators) in general (i.e. as being independent of sensibility), and thus things (initiators) in themselves, since appearances are mere *representations* of initiators according to the conditions that sensibility imposes on our references to the initiators (A26/B42 ff., A34B51 ff.).

To sum up, transcendental idealism holds that conditioned references are *ambiguous* in their objects. First, *outside* necessity contexts, they refer to initiators, *et al.*, which doesn't make the references any less subject to the conditions in question, but *within* such contexts, they only refer to initiators, *et al. as they are referred to outside* the necessity contexts through these conditioned references, that is, as *appearances*.

[6] Testing for the Adequacy of the Interpretation. The present book is predicated on the view, argued for in my previous book, that with respect to the question of the possibility of knowledge, Kant employs transcendental idealism only to account for the possibility of our *a priori* knowledge of objects, not empirical knowledge. Consequently, with respect to epistemology, he has no interest in a theory that postulates appearances apart from his endeavor to explain the possibility of knowledge that is *a priori.* It is essential to that account that initiators be represented as necessarily being appearances, that space and time necessarily belong to appearances, and that appearances exist in space and time necessarily ac-

cording to the categories. It is also essential to his idealism that initiators be conditioned by the understanding independently of their being conditioned by sensibility, if the initiators are to be represented as necessarily categorized, and thus be represented as necessarily being things (initiators) in themselves. These four propositions of *de re* necessity, *inter alia*, where the *alia* especially includes propositions about oneself, which are not treated in the present study, are constitutive of transcendental idealism, at least in its positive account of the possibility of *a priori* knowledge.

We can therefore test for the adequacy of our interpretation by determining the degree to which it succeeds as an interpretation of these four of Kant's uses of *de re* necessity. To do so, we will first attempt to make the interpretation more perspicuous by finding a suitable logical form for the expression of all four propositions we want to interpret. The next chapter finds such a form in a work of David Kaplan's, and the following chapter tests for the adequacy of the interpretation as it is represented through Kaplan's interpretation of our *linguistic* uses of *de re* necessity. It does so by applying Kaplan's interpretation to Kant's four uses of the necessity in a more detailed way than has been done, albeit implicitly, in this chapter. As indicated just above, the degree to which it is successful will be the extent to which the form renders our new epistemological interpretation of Kant's transcendental idealism more perspicuous than it might be without the formulation of the interpretation in logical terms. The added perspicuity is intended not just for Kant scholars, but for analytic philosophers generally, many of whom are probably relatively unversed in the Kantian system, but who, because of their familiarity with the (non- transcendental) logic that is used to express our Kaplanesque interpretation of Kant's uses of *de re* necessities, ought to recognize at least the logically salient features of the interpretation of Kant that is being offered here.

Chapter 7 – Kaplan's Referential Ambiguity

[1] Necessity and Quotation. David Kaplan disposes of Quine's skepticism concerning *de re* necessity through an ingenious interpretation of the quotation that is involved in the linguistic expression of the necessity. He argues that Quine is mistaken when he faults the necessity for the opacity of such contexts; it is rather due to the quotation. To avoid the opacity, Kaplan adopts what he considers Frege's employment of quotation in Fregean *indirect* discourse, which allows Kaplan to give the necessity a transparent interpretation. How Kaplan accomplishes this feat is the topic of this chapter.

For those who want to try to follow Quine and still get as close as possible to an expression of the idea of a necessary property or relation belonging to an object whose reference occurs in a purely *transparent* position, Kaplan commends Quine's *relational* sense of necessity, which Quine had introduced to express what he called "propositional attitudes."[1] For example, in Quine's relational sense of necessity, it can be said that

Nine is such that it is necessarily greater than five (114).[2]

Though this expression of the proposition we are interested in both captures essential features of our idea of *de re* necessity and yet allows the occurrence of 'nine' to occupy a purely referential position,[3] thereby allowing substitution of identities and existential generalization, Kaplan is un-

---

1 Quine, "Quantifiers and Propositional Attitudes," reprinted in *Reference and Modality*, Leonard Linsky, ed. (London: Oxford University Press, 1971), pp. 101–2.
To repeat, regarding my use of single and double quotation marks throughout what follows, double quotation marks will be used for terms, words, or expressions that are attributed or attributal to someone in particular, but when not so attributed or attributable, as for example, a term in general use, single quotation marks will be used.

2 Textual numerical notations in parentheses and *without* capital letters are references to Kaplan's "Quantifying In," *locus. cit.*

3 Thus, the operator has what is called "narrower scope."Again, for a discussion of the distinction between "narrower" and "wider" scope of a modality such as necessity, see, for example, Arthur F. Smullyan, "Modality and Description," *The Journal of Symbolic Logic*, 1948, Vol. 13, reprinted in L. Linsky, ed., *op. cit.*

happy with it. Since we do not find the form of the expression at all "natural," a special explanation of the form is required – something that an interpretation of *de re* necessity should not have to be encumbered with (115). To compound the problem, one might very well try to explain the sentence as expressing a condition that the number nine *must* satisfy, in contrast to conditions that it satisfies only contingently. And this would land us in the very metaphysics of traditional Aristotelian essentialism that Quine had warned against when pointing out the pitfalls of trying to quantify into necessity contexts.

Before leaving Kaplan's account of Quine, we should note that in "Quantifying In" Kaplan seems to accept uncritically Quine's view that quantifying into a necessity context in Quine's relational sense of necessity entails traditional Aristotelian essentialism. But as we have already discussed, a distinction can be drawn along Kit Fine's lines between the variables in such a quantification being open to substitution *salva veritate* and the *objects*' of such a quantification necessarily fulfilling a certain condition. The former is the logico-syntactic interpretation of the problem of quantifying in and the latter is the metaphysical problem of essentialism. It is only the latter about which Quine asserts that properties can be said to belong to objects *necessarily* only if the objects are considered as satisfying a certain condition or description, but not the objects themselves apart from such satisfaction or their being so described. To reiterate the position adopted in this book, however, it is only the former understanding of the problem of opacity that underlies the adaptation of Kaplan to an interpretation of Kant's uses of *de re* necessity. It is also the position of this book, however, that while Kant's transcendental idealism can be understood as a response to the mere logico-syntactic side of the problem of opacity, it nonetheless considers objects – initiators, *et al.* – *only* under certain, indicated conditions of representation, when it ascribes specific necessary properties and relations to these objects (appearance, space and time, and the categories), something that Quine says is necessary only for a *metaphysical* understanding of the problem. Finally, it should be noted that it is not clear that Kaplan is working with the distinction and has deliberately opted for the latter understanding.

Continuing with Kaplan, he asserts that as long as it is assumed, as he says Quine does assume, that expressions always denote what they usually denote, regardless of the context in which they occur, and specifically, where the context is quoted, the objects over which we can quantify into necessity contexts cannot be found. The reason Kaplan gives is that within a quoted necessity context an expression ought *not* be taken to denote what it usually denotes. Consequently, Kaplan recommends that we give up the assumption in question – that we cease to follow Quine's "fanatical monodenotational[ism]" – and follow Frege instead, who argues that in discourse he calls "indirect," which covers quoted contexts, expressions denote something *other* than what they usually denote (118). Frege took the objects of indirect discourse to be either his so-called senses (*Sinne*) of expressions, which were discussed above in chapter 5, or expressions that are usually used to make direct references to objects.

Kaplan therefore recommends that we follow Frege in our interpretation of the objects that are denoted within necessity contexts and, to be on the safe side ontologically speaking, that we let what are denoted within such contexts be *expressions* instead of Fregean senses, the latter being more than Kaplan wants to argue for in a paper designed to respond to Quine's concern about opacity and not to defend the ontological status of Fregean senses. The task before Kaplan, therefore, is to quantify over *expressions*, specifically those that denote objects, instead of quantifying over the *objects* denoted by the expressions, which is what Quine wants to do. In the case of our illustrative proposition, we should quantify over expressions, including numerals, and not numbers.

In further explanation of his interpretation of our uses of *de re* necessity, Kaplan stipulates that the range of expressions that are to be quantified over must be restricted in a certain way. Our illustrative proposition, for example, (i.e. that the number nine is necessarily greater than five) requires that the expressions in question *necessarily* denote the number nine. For example, the restriction rules out expressions that denote the number through *empirical* conditions, and are thus subject to "*empirical* vicissitudes," such as astronomical facts. It therefore excludes descriptions that are true of objects only contingently, such as 'the number of planets' (129).

Kaplan thinks that this restriction is connected to a further restriction. He believes that expressions can necessarily denote only *abstract* objects: Denotation of concrete objects will always be subject to the "empirical vicissitudes" just mentioned, which make denotation of concrete objects *contingent.* That is why Kaplan provides distinct interpretations of quantifying in for necessity contexts and perceptual contexts. The latter, since they involve "empirical vicissitudes," require an interpretation that accounts for such contingency.[4]

---

4 How necessary cognitions can *necessarily* refer to empirical intuitions and thus

Kaplan further stipulates that outside necessity contexts, expressions denote their usual, or customary, objects – in our case, the number nine. But within such contexts, it is expressions that are denoted. If we quantify over variables that take expressions and not objects as their values, the variables will therefore denote expressions. So, *expressions* are the quantified objects in Kaplan's interpretation, in the sense in which the expression 'quantified object' was used in the last chapter. Outside quantified necessity contexts, the relevant expressions will necessarily directly denote the objects they usually denote. But *within* the contexts, the expressions denote these same objects "only *indirectly* by way of denoting some intermediate entity," viz., themselves (118). In this manner, replacing Quine's "monodenotational[ism]," which, except for his relational sense of necessity, *blocks* quantification into necessity contexts, with Frege's studied *ambiguity* (between expressions' denoting their customary objects and their also denoting something else), which *allows* quantification into such contexts, Kaplan can show us how we can quantify into those contexts, and thus avoid inducing opacity.

As already noted in the Preface, that a philosopher such as Kaplan can use a Fregean approach to interpret expressions occurring both without and within necessity contexts, and hence giving an interpretation of *de re* necessity, is supported by independent arguments advanced by both Gareth Evans and John McDowell.[5] They maintain that Frege's adoption of his sense- (*Sinn*) reference (*Bedeutung*) distinction is no bar to a Fregean *de re* interpretation of (certain of) our ascriptions of attitudes or thoughts to objects *de re*. Since neither Evans nor McDowell seems to restrict his remarks about the permissibility of Fregean *de re* interpretations

---

their corresponding objects, i.e. appearances, is of course the central problem that is supposed to be solved by Kant's transcendental arguments – both the transcendental expositions of the concepts of space and time and the Transcendental Deduction of the categories.

5 See Evans's dispute with Michael Dummett concerning Dummett's *Frege: Philosophy of Language* (London: Duckworth, 1973), ch 12, cited by Evans in *The Varieties of Reference*, ed. by John McDowell (Oxford: Clarendon Press, 1982), pp. 12–13. See also, Evans, *op. cit.*, chaps. 1, 6, and 7, cited by John McDowell in *Meaning, Knowledge, and Reality* (Cambridge, MA: Harvard University Press, 1998) p. 214. And, on the same issue, McDowell disputes Tyler Burge's views that are presented in Burge's "Belief *De Re*," in *Journal of Philosophy* 77 (1974), cited by McDowell in "*De Re* Senses," *loc. cit.*, pp. 215 ff. Also see McDowell's own arguments in support of the idea of interpreting Frege's theory as allowing ascriptions of attitudes or thoughts to objects *de re*, pp. 214–27, and 261–74.

of such attitudes, their arguments can be used at least as a background in support of Kaplan's idea that specifically *de re necessity* contexts can be given a Fregean interpretation.

To sum up, *expressions are the quantified objects in Kaplan's interpretation.* Outside quantified necessity contexts, expressions directly denote their customary objects, which are abstract objects, and they do so necessarily. But within the contexts, the expressions again denote these same objects, but only *indirectly,* by way of their reference back to their occurrences outside the contexts. With regard to *quantifying over* denoting expressions, such as numerals, and thus quantifying into necessity contexts, the variables that are bound by the quantifiers will take the indicated denoting expressions as their values. In this manner, replacing Quine's "monodenotational[ism]," which, except for his relational sense of necessity, blocks quantification into necessity contexts, with Frege's studied ambiguity (between expressions' both directly denoting objects outside such contexts and their indirectly denoting the objects within the contexts through their references back to their occurrences outside the contexts), Kaplan can show us how we can quantify into necessity contexts.

[2] Kant's and Kaplan's Referential Ambiguity. As the previous chapter makes clear, I interpret Kant's transcendental idealism as employing *de re* necessity (of the logical or syntactical type, as explained by Kit Fine) which, like Kaplan's characterization of his own interpretation of our uses of *de re* necessity, I characterized in the same chapter as being *ambiguous* between two levels of discourse about objects: When Kant ascribes necessary properties or relations of appearance, space and time, and the categories to initiators, *et al.*, he takes conditioned references occurring outside necessity contexts as referring directly to initiators, *et al.* But within necessity contexts the references – what I have called in sections 3 and 4 of the last chapter the *quantified objects* – refer back to the antecedent conditioned references, and therefore refer to initiators, *et al.* only *indirectly* – through these antecedent references to the objects. His *de re* necessary discourse is thus about *both* the objects and the conditioned references to them. The necessity doesn't induce opacity, since it is not the initiators, *et al.* as they are referred to *independently* of the conditions on the references to them, but the conditioned references themselves that are quantified into the necessity contexts. Consequently, it is not initiators, *et al.* as *independent objects* to which the necessary properties and relations of appearances, space and time, and objects of the 'I think' are ascribed (and hence, the necessary properties and relations are not ascribed to ini-

tiators, or things, *in themselves*), but rather the initiators, *et al.* only *as they are referred to by the conditioned references.*

On Kaplan's interpretation of Frege, the latter would say of Kant's uses of *de re* necessity that his ascription of necessary properties and relations occurs in discourse that is both direct and indirect: Only *as they are referred to* by conditioned references can initiators occur in the proposition that they must be appearances; a similar point holds with respect to the conditioned references to appearances and ascribing to them the necessary properties and relations of space and time; and finally, a similar point holds with respect to the conditioned references to appearances in space and time and ascribing to them the necessary properties and relations of the categories. That Frege would attribute this trivial truism to Kant – that references refer to objects only as they are thus referred to, or only as objects of the references, where the objects of the references are *jointly* determined by the initiators, *et al.* and the conditions on the references – is supported precisely by the interpretation of Frege's philosophy of language that is advanced by Evans and McDowell as previously mentioned. Yet the initiators, *et al.*, to which appearances, space and time, and the categories are ascribed, are referred to *outside* the necessity operator precisely as the objects about which Kant says in *direct discourse* that they are referred to by means of sensibility, outer and inner sense, and the 'I think.' So, according to Kaplan's notion of Fregean ambiguity, Kant's *conditioned references* are referentially *ambiguous: Outside* necessity contexts, they refer *directly* to what they usually refer to, viz., objects that are referred to independently of the conditions on these very references themselves, viz., initiators, *et al.*; yet, at the same time, the conditioned references nonetheless directly refer to them. But *within* the necessity contexts, since they are the quantified objects, the conditioned references refer to the original objects – initiators, appearances, and appearances in space and time, respectively – only *indirectly*, by referring back to their antecedent occurrences which are *outside* the necessity contexts and which refer to the objects *directly.* Consequently, they do not refer to the objects independently of the conditions that limit them. Within the necessity contexts, initiators, *et al.* simply *cannot* be referred to independently of these conditions: They can be referred to only *conditionally.* That is why opacity is not induced.

[3] The Formulation of Kaplan's Interpretation of *De Re* Necessity. Again, the proposition of *de re* necessity Kaplan wants to interpret – our illustrative proposition – is

(1) Nine is necessarily greater than five

which in Quine's relational sense of necessity appears as

(2) Nine is such that necessarily it is greater than five

To get his own interpretation, Kaplan begins with a proposition that involves quotation:

(3) Necessarily 'nine is greater than five' (122).

The interpretation he eventually arrives at takes four steps. First, as already noted, he follows Frege and allows denoting expressions within quotations to denote something other than what they usually denote. In (3) 'nine' inside the quotation denotes itself instead of the number nine (122).

A problem immediately arises, however. In (1) the numeral 'nine' denotes the number nine, and not itself. How can Kaplan require the quantifier to bind over expressions and not the objects (e.g. the numbers) denoted, as a generalization from (3) would entail, when what is actually denoted in (1) and (3) is an object – a number – and not a numeral, or more generally, not a linguistic expression? The second step of Kaplan's procedure answers this question. He places the quotation within the scope of what, borrowing from Church, he calls a "denotation operator." According to it, a denoting expression such as the numeral 'nine' denotes an object, such as the number nine.

Let us look at Kaplan's first approximation of his interpretation:

(4) ∃α(**Δ** (α, nine) & **N** 'α is greater than five') (123)

The existential quantifier binds over the variable, represented by the lower case Greek letter 'α', which takes linguistic expressions as values. '**N**' stands for necessity. The predicate constant within the quotation, 'is greater than five', denotes itself. Church's denotation operator, standing for the relation of denotation between an expression and an object, is represented by the upper case Greek letter '**Δ**'. And the object Kaplan is concerned to denote, the number, indeed *is* denoted by the use of the numeral 'nine' as it occurs within the scope of the denotation operator, whereas the variable, both inside and outside the quotes, takes an expression and not a number as a value.

The third step employs Quine's "corner quotes" ('⌜ ⌝') instead of regular quotes in order to accommodate this Frege-Church interpretation of the necessity (120).[6] Corner quotes have the effect of quoting only the *context* in which substitutable expressions occur, which in our case includes the predicate constant, but not the variable. They thereby allow the variable to take the same expression as a value both inside and outside the quotes. Kaplan claims this is "Frege's view of quotation contexts . . ." and dubs them accordingly, "Frege quotes." For Kaplan, they signal possible quantification into necessity contexts from without. That a bound variable can take linguistic expressions as values both outside a necessary proposition and within it is the main idea behind Kaplan's Fregean interpretation of *de re* necessity.

The fourth step bars any expression that introduces contingency from being taken as a value for the variable. In the present case, only expressions that *necessarily* denote the number nine are allowed as values for the variable. The numeral 'nine' and the expression 'three times three' belong to the range of values for the variable, but 'the number of planets' does not. Kaplan's use of *necessary denotation* makes a virtue out of Quine's objection that "an invidious attitude toward certain ways of uniquely specifying" an object is a ". . . reversion to Aristotelian essentialism . . ."[7] Kaplan freely admits to having such an attitude toward specifying an object without admitting to the "metaphysical tradition connected with" Aristotelian essentialism (130).

This provides the interpretation of (3) that Kaplan wants:

(5) $\exists\alpha(\mathbf{\Delta}_N(\alpha, \text{nine})\ \&\ \mathbf{N}\ulcorner\alpha \text{ is greater than five}\urcorner)$ (130).

It can in turn be used to interpret (1), by omitting the quantifier and substituting 'nine' for the variable.

(6) $\mathbf{\Delta}_N(\text{'nine'}, \text{nine})\ \&\ \mathbf{N}\ulcorner\text{nine is greater than five}\urcorner)$.

Kaplan's essentialism follows Church instead of Aristotle (131). As already noted in the first section of this chapter, Kaplan maintains that the relationship of necessary denotation between an expression and an object holds only if the relationship is "fixed on logical, or perhaps . . . linguistic, grounds alone" (128). Empirical vicissitudes such as those that determine how many planets exist would undo the necessity that might otherwise attach to the denotation. Therefore, Kaplan concludes,

6 Kaplan coyly acknowledges a technical and typographical debt to Quine, who introduced "corners," "corner-quotes," or "*quasi-quotation*" (Ibid., 120, fn. 9). See W.V. O. Quine, *Mathematical Logic* (New York: W.W. Norton, 1940), p. 33.

7 Quine, "Reference and Modality," p. 30.

the only objects that could be denoted on these limited grounds are abstract objects, such as numbers – a far cry from the "metaphysical tradition connected with Aristotelian essentialism."

Our next job, obviously, is to determine the extent to which we can apply these results we have now got from Kaplan to our interpretation of Kant. Though Kaplan on quantifying in, or on anything else for that matter, is not recognized by any Kant scholar as an arbiter among competing interpretations of Kant's transcendental idealism, as a perspicuous formula for interpreting our own uses of *de re* necessity, the perspicuity of the formula – as perspicuous as it is – alone counts in favor of using it as a measure of the success of our own interpretation of Kant. That it dovetails so nicely in so many respects with Kant's uses of the necessity, far from being a mark against it on grounds of circularity, actually counts in its favor, for the perspicuity of Kaplan's interpretation can be determined on grounds that are completely independent of Kant, even though the Fregean tradition within which he is working is an identifiable legacy of Kant's theory of the relation between objects and their necessary representations.

[4] Kant's and Frege's Referential Ambiguity. From the last chapter it has been clear that I interpret Kant's transcendental idealism as employing *de re* necessity (of the logico-syntactic type) in a manner that, like Kaplan's characterization of his own interpretation of *de re* necessity, can be characterized as employing an *ambiguity* between two levels of discourse about objects: When Kant ascribes necessary properties or relations of appearance, space and time, and the categories to initiators, *et al.*, he takes conditioned references occurring *outside* necessity contexts as referring to initiators, *et al.* But as occurring *within* necessity contexts, they refer to initiators, *et al.* only as they are referred to by these conditioned references that occur outside the necessity contexts. His discourse is thus about *both* empirical intuitions and initiators, which in affecting us are given to us through the intuitions – to repeat our identification of initiators that was introduced in chapter 2. But it is still the initiators, yet only as they are represented through their appearances, to which he is ascribing the necessary properties and relations of space, time, and the categories – insofar as the categories determine their order in space and time. Quite apart from Kant, as we shall presently see, Kaplan says the same thing in regard to his interpretation of the ascription of necessary properties to the number nine: The ascription is indeed interpretable, but only if the expression denoting the number denotes it in a prescribed manner, namely, that the denotation is *necessary.* This effectively excludes

'the number of planets' from the denotation. Yet, it must be immediately acknowledged, that where Kant's references to initiators, *et al.* are *conditioned*, Kaplan's are not, since necessarily denoting objects is not a condition on the expressions, at least not in the sense in which Kant's references to initiators, *et al.* are conditioned by sensibility, outer and inner sense, or the 'I think'.

On Kaplan's interpretation of *de re* necessity, Frege would say that *Kant's* ascription of these properties and relations occurs in discourse that is only indirect: Only as they are represented through our sensible representations *as appearances* can initiators occur in the scope of the necessity operator in a proposition whose truth we can know. Yet the initiators that the appearances represent are denoted outside the operator precisely as the objects about which it is said in *direct discourse* that they are represented through the representations in question. So, according to Kaplan's notion of Fregean referential ambiguity, Kant's use of the concept of *appearance* in a proposition is referentially *ambiguous:* It denotes both appearances and the initiators they represent. At the same time, the use crosses the two levels of discourse – direct and indirect. Indeed, it can be said that the crossing is nothing more than the Fregean referential ambiguity. Or, perhaps less anachronistically, we should say that Frege's crossing of these two levels of discourse – in any case, as Kaplan interprets him – follows Kant's uses of referential ambiguity. The noticeable degree of concordance between our interpretation of Kant and Kaplan's Frege thus augurs well for the test of the adequacy of Kaplan's formula expressed in logical terms when measured against our epistemological interpretation of Kant's transcendental idealism, which is carried out in the next chapter. The coincidence of the crossing of levels and the Fregean ambiguity indicates that Kaplan's formula will indeed provide a template that ought to make our independent but co-extensive epistemological interpretation of Kant's idealism more perspicuous.

Chapter 8 – Kaplan's Interpretation Adapted to Kant

I Introduction

[1] The Order of Kant's Uses of *De Re* Necessity. We pick up our discussion of Kant with a review of the order of Kant's uses of *de re* necessity. It begins with the claim of dual-aspect interpreters that for Kant the possibility of knowledge, and I would add, the possibility of specifically *a priori* knowledge, can best be understood on the supposition that we can determine a *single* set of objects in two ways – by adopting what has come to be called a *dual-aspect* interpretation of transcendental idealism. According to the epistemological criterion of Kant's ontology offered in this book, moreover, the identical objects in question are objects that in affecting us are given to us. In line with the transcendentally logical criterion of Kant's ontology of my previous book, I have followed Kant and consider these identical objects Kant's (basic) reality and, because they are the (external) objects that get our knowledge started, I have called them *initiators.* Therefore, initiators are the objects that in affecting us are given to us.

The order of Kant's uses of *de re* necessity can readily be divided into nine steps, largely, but not completely, following the order of the *Critique,* including the Transcendental Deduction of the categories. First, and beginning with the Aesthetic, initiators must be represented through (as) *appearances,* if they are represented by means of sensibility. Second, the appearances must be determined (more generally, represented) through space and/or time, if they are represented by means of outer and/or inner sense, respectively. Third, introducing the Logic, sensibility is abstracted from the overall argument and therefore appearances in space and time are abstracted as well. What is left are objects of a sensible intuition *in general,* which must be determined *intellectually* (more generally, represented) through the categories, if they are represented by means of the 'I think', (and hence, be "objects for the subject" [B132]). Fourth, appearances (from the Aesthetic) must be determined *a priori imaginatively* ("productively") through the categories, since they must be objects of inner sense (all appearances are such objects, [B150, A34/B50]). To be "determined" here means that inner sense must be "affected" by the understanding (through the categories), and objects of inner sense, which includes all appearances as objects that can be given to us, are thus necessarily determined through the categories. Fifth, again from the Aesthetic, empirical intuitions are reintroduced into the argument, and their manifolds must be represented as objects of perception, if they are connected (represented) by means of the empirical imagination, designated as the *synthesis of apprehension* (B160). These connections make *perceptions* possible. Sixth, since the forms of outer and inner sense – space and time – are themselves (*a priori*) *intuitions*, they, or their respective manifolds, must be determined *intellectually* through the categories, if they are represented by means of the 'I think'. Seventh, again from the Aesthetic, appearances in space and time are reintroduced into the argument, and their respective manifolds must be intellectually determined in space or in time, through the categories, if they are represented by means of the 'I think'. Eighth, the manifold of *perceptions* in space and time must be connected through the categories as a possible *experience*, or *empirical knowledge*, if they are represented by the *a priori* ("productive") imagination, which carries the 'I think' with it. Ninth, and finally, initiators must be represented *only* intellectually through the categories as *things in themselves*, if they are represented by means of the 'I think' apart from sensibility.

The adaptation of Kaplan's interpretation of our uses of *de re* necessity to Kant's uses of the same necessity, however, will cover only *four* of the steps that occur in the order of Kant's overall argument given just above. These will be the *first*, *second*, *seventh*, and *ninth* steps just mentioned. If these adaptations of Kaplan's interpretation to Kant's uses have been made clear, the other steps can be taken by the reader him/ or herself, if so desired.

First, a terminological note: Though I would prefer to use the term 'reference' for the relation Kant typically has in mind when he uses the German '*Beziehung*' for the relation of knowledge or representation to objects, for the sake of conforming to Kaplan's interpretation of *de re* necessity, in this chapter, where I am adapting his interpretation of the *de re* necessity that is independent of Aristotelian essentialism to Kant's use of the necessity, I will follow Kaplan and use 'denotation' for the reference relation.

Second, it is crucial to my interpretation of Kant's uses of *de re* necessity that we understand the denotation of objects according to the relation of presupposition developed in chapter 5 above (and in chapter 9 below). Accordingly, the denotation presupposes, and does not entail, the existence of the object denoted.

Also, although appearances are objects of *either* our understanding *and* sensibility *or* sensibility alone, Kant actually introduces them by way of sensibility alone (A20/B34). Accordingly, I will follow the actual course of Kant's uses of *de re* necessity of concern to us and divide the uses according to his two means of representation, sensibility and the understanding. An interpretation of the uses that involve the understanding will be given in Part III of the chapter, in sections 5 and 6, below. Regarding the uses that involve sensibility, they can now be further divided.

[2] The First Division of Uses of *De Re* Necessity Involving Sensibility. The first division of Kant's uses of *de re* necessity involving sensibility consists in Kant's statement that initiators must be represented as appearances, if they are denoted through a mode of sensibility (B68, cf. A42/B59, A49/B66). That is, if they stand under the condition of sensibility, initiators must be represented as appearances. In (1) below he is stating of initiators that are (contingently) denoted through sensibility[1] that they must be appearances, and he can make that statement without thereby, i. e. in the very making of the statement, denoting them through sensibility. This can be done, since he himself is denoting them in what, according to Kaplan's account that was given in the previous chapter, Frege would call *direct discourse* – which, in this case, is epistemological discourse – and, more importantly, only *mentioning*, but not *using*, or engaging, sensibility in doing so.

Before we can interpret the first in our series of Kant's four uses of *de re* necessity of concern to us on Kaplan's model, we must replace the linguistic *expressions* that instantiated Kaplan's use of the existential quantifier with Kant's mental *representations.* Otherwise, we won't be able to adapt Kaplan to our epistemological criterion of Kant's ontology. This switch from linguistic expressions to mental representations has a certain consequence which needs to be mentioned.

Since we are about to adapt Kaplan's use of the existential quantifier to our own use of it in quantifying over certain of Kant's *representations*, we must withhold the use of the quantifier about to be attributed to Kant from being measured against Kant's criterion of existence in general that was given in chapter 3 above. In other words, the notion of existence that is employed in Kaplan's use of the existential quantifier to bind over *linguistic expressions* and now our use of it to bind over Kant's *mental representations* cannot be held to conform to the criterion of existence in chap-

1 The denotation is contingent because it is a contingent fact that initiators affect us.

ter 3, since the Kantian representations that are to figure in the account of Kant's uses of *de re* necessity that follows (along Kaplan's lines) cannot be said to exist according to that criterion. In short, these particular representations are not objects that can be given through intuition. It is not just concepts and objects involving syntheses of manifolds that cannot be given through intuition, but intuitions themselves are not objects that can be given through further intuitions. As we saw in chapter 4, for Kant, either we empirically intuit *initiators* or we intuit *a priori* their *a priori determinations* – space and time, but we cannot intuit, either empirically or *a priori*, intuitions themselves.

Though certain of Kant's *sensible representations* do indeed satisfy the criterion of existence in chapter 3, for example, appearances do, nonetheless, just as it makes no sense even to try to apply the criterion to Kaplan's *linguistic expressions*, so we would confound our Kaplanesque account of Kant's uses of *de re* necessity if we tried to use the criterion to determine the representations that can instantiate the variable in Kant's putative employment of the necessity in those uses. So our conclusion is that we must withhold the criterion of existence that was given in chapter 3 from our present Kaplanesque use of the existential quantifier in binding over Kant's mental representations in our interpretation of Kant's uses of *de re* necessity that are considered immediately below.

II Sensibility

The uses of sensibility can themselves be divided according to whether initiators are represented as appearances or appearances are represented as existing in space and time.

[3] Initiators as Appearances. We can now interpret the first in our series of Kant's uses of *de re* necessity on Kaplan's model. In (1) below instantiate the variable with Kant's *mental representations* that are given by means of *sensibility*; signify this constraint by the letter 'S' in the denotation operator; let the quotes be what Kaplan calls "Frege quotes"[2]; and let the representations inside the quotes, except for the variable, denote themselves; let '**N**' stand for necessity; and outside the quotes employ the (merely contingent) denotation operator. So (1) would be our

2 See the last chapter, section 3, for an explanation of Kaplan's calling the quotes "Frege quotes."

Kaplanesque interpretation of Kant's necessary ascription of *appearance* to initiators:

(1) $\exists\alpha$ ($\mathbf{\Delta}_S$ (α, an initiator) & $\mathbf{N}\ulcorner\alpha$ is *appearance*$\urcorner$).

[4] Appearances as Spatial-Temporal Objects. Kant's Transcendental Expositions of the concepts of space and time in the *Critique* (§§3 & 5) entail that they necessarily belong to appearances (A38/B55). However, his preceding Metaphysical Expositions of the concepts, respectively, do not warrant that entailment, since they only establish that space and time are *a priori* intuitions (§§2 & 4). To get the entailment, Kant first needs the premise that necessarily, these intuitions are space and time – a logical conversion by limitation of the proposition that necessarily, space and time are *a priori* intuitions. If he can identify a particular *a priori* intuition as space and another as time, he *could* infer that our *a priori* intuitions indeed *are* identical with space and time, respectively. And second, the entailment requires the premise that our *a priori* intuitions necessarily denote appearances. On the basis of both premises and the Metaphysical Expositions, he could then conclude the Transcendental Expositions with the *de re* proposition that appearances are necessarily spatial and temporal.

In regard to the first premise, Kant had already maintained that by means of outer sense we represent objects as outside us spatially and by means of inner sense we represent objects as inside the mind – as states of mind – temporally (A22–3/B37–8). If we now add these connections between space and outer sense and time and inner sense, respectively, to the proposition that space and time are *a priori* intuitions, we can get the conclusion that our outer *a priori* intuition *is* space and our inner *a priori* intuition is time. Within Kant's system, since both the premises of the Metaphysical Expositions and the connections between space and outer sense and time and inner sense, respectively, are *a priori*, the conclusion that our outer *a priori* intuition is space and our inner *a priori* intuition is time must also be *a priori*. Since the connections are *a priori*, they constitute facts about our faculties of knowledge.[3] Consequently, propositions expressing such facts belong to our knowledge of the mind that experience cannot controvert. In that sense, the truths expressing such facts are *a priori*, and accordingly, are necessary.

With respect to the premise that our *a priori* intuitions necessarily denote appearances, Kant explicitly argues, in regard to space alone, that

3 See chapter 2, section 2, above.

since outer intuition "precedes the objects themselves," it can determine *a priori* "the concept of these objects" (A26/B41). I take it that Kant means that in determining the concept of outer appearances, outer intuition determines the appearances. Similar reasoning can be applied to time. Since inner intuition helps make experience possible, it can determine the concept of experience, and thus determine the concept of objects of possible experience – again, appearances – and thus determine the appearances.

When these two premises are conjoined with the Metaphysical Expositions, Kant can make the following argument for the proposition that appearances are necessarily spatial and temporal. Since the concept of appearances that can be given to us is determined by our outer and inner *a priori* intuitions, and since it is *a priori* necessary that those intuitions are space and time, respectively, the concept of appearances that can be given to us is determined by space and time. Moreover, I assume that "the concept of appearances" means "what appearances are thought or understood to be," and since space and time, as the determinations of our outer and inner *a priori* intuitions, determine the concept of appearances, they determine what we understand appearances to be. Consequently, Kant is entitled to conclude that space and time "necessarily ... [belong] to the appearances of this [independent] object" (A38/B55). This conclusion is the second of Kant's uses of *de re* necessity that we are going to interpret according to Kaplan's model.

Before I offer my Kaplanesque interpretation of the proposition in question, however, lest it be objected that in what follows I am ignoring an essential part of Kant's theory, I should state that I am aware that he considers the spatial and temporal determinations of appearances by outer and inner sense to be as distinct from each other as are space and time themselves, and also as are outer and inner sense themselves. But though that distinction is essential to an account of Kant's theory of the possibility of *a priori* knowledge, it seems immaterial to an interpretation of his use of *de re* necessity in stating that space and time necessarily belong to appearances.

Our interpretation of the proposition just stated can now proceed à la Kaplan as follows:

(2) $\exists\alpha\ (\mathbf{\Delta}(_{\text{O \& I}),\,\text{N}}\ (\alpha, \text{an appearance})\ \&\ \mathbf{N}\ulcorner\alpha \text{ is spatial and temporal}\urcorner)$

To read the interpretation, instantiate the variable with mental representations; let them be determined by outer and inner sense, signified by 'O & I' and let them denote appearances; let the quotes be "Frege quotes" and let the representations inside the quotes, except for the variable, denote themselves; let '**N**' stand for necessity; and employ the necessary denotation operator for the relation of necessary denotation between our representations of outer and inner sense, on the one hand, and appearances, on the other.

III The Understanding

We now turn to Kant's uses of *de re* necessity that involve the understanding. Like the uses for sensibility, these can be divided further.

[5] Appearances as Categorized. The first division consists of Kant's statement that spatial and temporal appearances must conform to the categories, which is how he invites us to understand the proposition that nature must conform to law (B159–60, B164–65).

The proposition whose *de re* necessity we are now interpreting – that such appearances must conform to the categories – is nothing less than the very thesis that the Transcendental Deduction of the categories is supposed to prove (A90/B122). The problem for the Deduction is that appearances in space and time *might not* be subject to the categories (A90/B123). The Deduction would fail unless this possibility were bypassed.

The possibility itself is a consequence of the dependence of these appearances on sensibility alone[4] and the mutual independence of sensibility and the understanding. It therefore allows the appearances to be given to us by means of sensibility independently of the understanding. It thus allows the appearances to be in a state of "confusion," thereby leaving the categories without any object that may be given to us, i.e. empty (A90–91/B123).

In the interpretation that follows, let pure concepts of the understanding, or categories, be the representations that denote spatial and temporal appearances by means of the understanding (A79/B105). If they do so only contingently, however, it will be merely contingent that the appearances conform to categories. Although the appearances may thereby indeed be denoted by categories, that will not solve the problem of the Deduction, since the contingency of their denotation allows that they might be in a state of confusion, the very possibility the Deduction is supposed to bypass. A Transcendental Deduction is irrelevant to a ques-

4 Since it is not necessary to argue the point here, I am assuming that appearances are due to sensibility alone, even though the objects of experience, which involves both sensibility and the understanding, are also appearances. I argue the point in *Kant's Theory*, chs.7 and 8.

tion of fact – "*quid facti*" – namely, whether these appearances in fact conform to categories (A84/B116, Cf. A87/B119). Rather, it must show that the appearances *must* conform to categories (A90/B123). That alone can answer "the question of the right (*quid juris*)" to employ categories (A84/B116): Their only *justification* is that they *must* be so used.

In the Metaphysical Deduction of the categories (A76/B102 ff.),[5] Kant moves from categories *in general* to the *particular* categories that are contained in the Table of Categories – *the* categories – such as substance and causality. He does so by using the logical functions of judgment to form particular categories out of categories in general. Each category that results has its own logical function of judgment.[6] Since each category is assigned exactly one function (A79/B105), Kant can characterize a category as a determination of "the intuition of an object . . . in respect of one of the logical functions of judgment" (B128). Since the premises of the Metaphysical Deduction are *a priori*, Kant can conclude that it is *de re* necessary that categories that denote the appearances are the particular categories listed in the Table of Categories.

If we adapt Kaplan to our interpretation of the Metaphysical Deduction we get:

(3) $\mathbf{N}\ulcorner\alpha$ is substantial or causal, etc.$\urcorner$

where we are to understand it as before, except that the mental representations of the appearances that can be values for the variable are the categories, not intuitions.

The new representations are again determined by constraints on the denotation operator, only now the understanding replaces the constraint of sensibility and, on the basis of (2) above, the appearances are qualified as being spatial and temporal. However, where our outer and inner *a priori* intuitions could determine appearances by determining their very concept, categories assigned a logical function of judgment cannot determine the appearances that way. The very contingency that creates the need for a

5 We continue with the part of our interpretation that represents the Metaphysical Deduction of the categories just as our interpretation of his use of *de re* necessity involving the concepts of space and time began with the part of our interpretation representing the *Metaphysical* Exposition of those concepts.

6 We recall that he made a similar move when he formed outer and inner *a priori* intuitions from outer and inner sense and then connected them with space and time, respectively, in the Aesthetic.

Transcendental Deduction – the logical independence of sensibility from the understanding – makes that impossible.

Categories *can* determine the appearances, however, in respect of the logical functions of judgment, since they can determine the *intuitions* of the appearances in that respect. They *must* so determine intuitions, and hence, their appearances, moreover, if the manifolds of the intuitions are to be objects for the subject, that is, if they are to be brought to the self-consciousness of the subject, since the categories are necessary conditions of that possibility (B143). Possible self-consciousness thus plays for Kant the role that "logical . . . or linguistic grounds" play for Kaplan in *his* explanation of the necessity with which an expression can denote an object.[7] Furthermore, if the intuitions are *a priori*, the appearances must "be represented as determined in space or in time" (B161). It should be noted here that, in light of what was said above in section 4, the distinctive determinations of appearances by space and time is more than "immaterial," as it was characterized there. Still, it does not affect the adaptation of Kaplan to Kant; it only makes it a bit more complicated. On the other hand, if the intuitions are empirical and belong to experience, the categories must determine the appearances as objects of experience (B128, Cf. B162, B165). Here, it would seem, appearances must be considered spatial *and* temporal, since their "sum" is said to constitute "nature" (B163), which involves both determinations of appearances – space and time – together.

Having completed our interpretation of the thesis the Transcendental Deduction is supposed to prove, we can adapt Kaplan's interpretation of *de re* necessity to our interpretation of Kant's use of it in the Transcendental Deduction:

(4) $\exists\alpha(\mathbf{\Delta}_{U,N}$ (α, a spatial and temporal appearance) & $\mathbf{N}\ulcorner\alpha$ is substantial or causal, etc.$\urcorner$)

The interpretation is to be understood as before, except, now, the categories are the possible values for the variable; appearances are qualified as spatial and temporal objects; and the letter "U" stands for the understanding. The result is an interpretation of Kant's ascription of the categories as necessary determinations of spatial and temporal objects. This is the third instance of Kant's use of *de re* necessity that is being interpreted according to Kaplan's formula.

7 See the last chapter, section 3.

[5] Initiators as Things in Themselves. There remain for our discussion two other of Kant's uses of *de re* necessity that involve the understanding. Both involve his notion of a thing in itself as an object that is independent of sensibility, and hence an object of only an understanding, but an understanding such as ours, not an intuitive understanding, such as that which belongs to a divine being (B145).[8]

Since the understanding represents objects as objects in general, which entails that it represents them as independent of sensibility, it must be able to represent initiators as independent of sensibility. Since objects that are represented as independent of sensibility are represented as things in themselves, we can arrive at an interpretation of Kant's statement that initiators that are objects of the understanding *apart from sensibility* must be Kant's *things in themselves.* On the model of Kaplan's interpretation, we thus have:

(5) $\exists\alpha(\boldsymbol{\Delta}_{U,N}$ (α, an initiator) & $\mathbf{N}\ulcorner\alpha$ is *in itself*$\urcorner$).

(5) should be read as the *correlate* of (1) above (cf. A30/B45).

[6] Things in Themselves as Categorized. Finally, since it is through the categories that the understanding represents objects, and since things in themselves are objects of the understanding (alone), they are necessarily subject to the categories (e.g. B309). Kant states that the understanding thinks things in themselves through what he calls *ideas* or concepts of reason (A320/B377). Once again using Kaplan's interpretation as a model, we can get:

(6) $\exists\alpha(\boldsymbol{\Delta}_{R,N}$ (α, a thing in itself) & $\mathbf{N}\ulcorner\alpha$ is substantial, causal, etc.$\urcorner$)

where the formula is to be understood as before, except that now the denotation is constrained by reason, signified in the formula by 'R'. (6) should be read as the correlate of (2) above.

[7] Conclusion. If we have been successful in modeling Kant on Kaplan, we might remark on Frege's relation to Kant: The extent to which Kaplan's interpretation depends on Frege and the degree to which it can be used to interpret Kant perhaps tells us that Frege was even closer to Kant than he at times acknowledged.

---

8 For discussion of that notion see chapter 3 above.

Chapter 9 – Geometry and Causality

I Geometry

[1] The Problem of *De Re* Necessity from the Subjective Standpoint. In Chapter 1 I said that certain of Kant's uses of *de re* necessity have been challenged by analytic philosophers from two opposite directions: The uses are under attack both objectively and subjectively. Quine's attack was from the side of objects: Our various manners of referring to objects entail that we prefer some manner(s) of reference over others, if we want to employ *de re* necessity; but such a preference leads directly to the tradition associated with Aristotelian essentialism. In the chapter before the last, Kaplan provided the basis of our Kantian response to Quine's skepticism, and in the last chapter the response itself was actually given.

Now we turn our attention to the challenge from "within," as it were. In this first part of the chapter we will meet the challenge to Kant's theory of the necessity of geometry that has been mounted by first, Russell, and then, Van Cleve. Russell's objection as resuscitated by Van Cleve is that since it is a contingent fact that we have the particular cognitive constitution that we in fact have, and since the truths of Euclidean geometry originate in that constitution, the truths can be no more necessary than the constitution; that is, they, too, must be merely contingent. And if these truths are contingent, so must be the Euclidean geometric properties and relations they express that are ascribed to appearances. Hence, there can be no *de re* necessity with which these properties and relations belong to appearances: A cognitive constitution different from ours would be a source of different, non-Euclidean geometric properties and relations belonging to appearances. And, *mutatis mutandis*, the same would be true of temporal relations that are ascribed to appearances: They cannot be necessary relations of appearances, either. Unless this challenge is successfully met, Kant's effort to explain the possibility of our knowledge of such *de re* necessities is quixotic, since, on Kant's own account of their origin, it would be directed at a mere chimera. The first purpose of this chapter, then, is to demonstrate that it is the attack from Russell and Van Cleve itself, and not Kant's explanation of our knowledge of such necessities, that unfortunately distorts its object. Second, the chapter will defend Kant against another challenge from "within." We will consider a situation in which our cognitive constitution is such that we do *not* think any objective succession of occurrences causally. This defense will be given in the second part of the chapter.

[2] The Russell-Van Cleve Statement of the Problem. James Van Cleve in his book on Kant's first *Critique* credits Russell with having delivered the *coup de grâce* to Kant's attempt to explain our attribution of necessity to certain of our propositions about objects.[1] For example, Kant's theory is supposed to explain the necessity that "we can only apprehend cubes as being eight-cornered,"[2] or that "2 + 2 = 4."[3] The explanation consists in his so-called Copernican Revolution in philosophy. It is our "cognitive constitution,"[4] specifically, our form of intuition, that explains our use of necessity in our talk about our apprehension of the number of corners of a cube or in our attribution of necessity to our addition of 2 plus 2 being equal to 4. But Russell points out that since our nature might have been different, and hence, Van Cleve continues, since "it is contingent we have the constitution we do . . . [then] if our constitution had been different, [the laws of arithmetic and geometry] would have been false, and other laws would have held their place."[5]

Van Cleve's formulation of Russell's argument consists first of a premise that claims that the necessity of any given proposition is explained by the fact that the proposition is "a deliverance of our form of intuition."[6] Van Cleve calls this the Dependency Premise, and using the square '$\Box$' for necessity, the diamond '$\Diamond$' for possibility, the single-line arrow '$\rightarrow$' for material implication, the double-line arrow '$\Rightarrow$' for strict implication, or entailment, and 'Fp' for "a proposition p is a deliverance of a form of intuition," he symbolizes this first premise of Russell's argument as follows:

1. $(p)(\Box p \rightarrow [\Box p \Rightarrow Fp])$

Since it is contingent that "any given proposition *is* delivered by our form of intuition," it is "possibly false" that it is delivered by our form of intuition; thus we get the second premise of Russell's argument[7]:

---

1 James Van Cleve and Bertrand Russell, *loc. cit.*, respectively, in Chapter 1.
2 Van Cleve, ibid.
3 Ibid., p. 40.
4 Ibid., p. 38.
5 Ibid.
6 ibid
7 Ibid.

2. (p) $\Diamond \sim Fp$

Van Cleve calls this the Contingency Premise. Moreover, if we let p be any necessary truth, we get the third premise:

3. $\Box p$

And from 3 and 1 we get the fourth proposition of the argument

4. $\Box p \Rightarrow Fp$

In addition, a theorem in modal logic holds that if two propositions are related by entailment, and the entailed proposition is possibly false, so is the proposition that entails it. So, from 2 and 4 we get:

5. $\Diamond \sim \Box p$

That is, it is possibly false that p is a necessary truth. Hence, under different conditions from those that actually exist the laws of arithmetic and geometry would be false. But this is equivalent to proposition 6 of the argument:

6. $\sim \Box\Box p$

That is, it is false that p is necessarily necessary. Hence, under different conditions from those that actually exist, viz., if our form of intuition were different from what it actually is, the laws of arithmetic and geometry would be false.

Van Cleve acknowledges that the defender of Kant might accept this conclusion, since it allows p to be necessary; it only rules out that it *must* be necessary: it would be merely "contingently necessary," Van Cleve says.[8] He cites Nicholas Rescher as having mounted such a defense.[9]

Although Van Cleve considers the conclusion about the contingency of the laws of arithmetic and geometry "absurd," as attested by an axiom of the modal system S4 that Van Cleve accepts, viz., $\Box p \rightarrow \Box\Box p$, he nonetheless goes after the Kantian who would exclude the axiom. He does so by strengthening the argument he has just given. Now the mere truth of a proposition instead of its necessity is what is supposed

8 Ibid., p. 39.

9 Nicholas Rescher, "Kant and the 'Special Constitution' of Man's Mind: The Ultimately Factual Basis of the Necessity and Universality of *A Priori* Synthetic Truths in Kant's Critical Philosophy," in *Studies in Modality*, American Philosophical Quarterly Monograph Series, 8 (Oxford: Oxford University Press, 1974), pp. 71–83, cited by Van Cleve, ibid., p. 38.

to depend on our form of intuition. He reasons that if a proposition owes its necessity to our cognitive constitution, so should its truth. The strengthened argument would thus be immune to the Kantian defense that Kant needs only a contingent necessity instead of a necessary necessity. The strengthened argument is rather concerned with the truth of p, not its necessity, and hence it can disregard the issue of whether the necessity in question is itself contingent or necessary. The strengthened argument involving only the truth of p goes as follows:

1. $(p)(\Box p \rightarrow [p \Rightarrow Fp])$ (Strengthened Dependency Premise)
2. $(p) \Diamond \sim Fp$ (Contingency Premise)
3. $\Box p$ (assumption for reductio)
4. $p \Rightarrow Fp$ (from 3 and 1)
5. $\Diamond \sim p$ (from 4, 2, and the same modal principle as before)
6. $\sim\Box p$ (from 5)[10]

The strengthened argument thus demonstrates that if our cognitive constitution were a logically necessary condition of the truth of p, as it ought to be if it were a logically necessary condition of the necessity of p, the assumption that p is necessary would lead to the absurd conclusion that it is not necessary. So Russell is vindicated after all: "If our nature changed drastically enough, we could wake up tomorrow and find that cubes have nine corners or that $2 + 2 = 5$."[11]

[3] The Problem with the Statement. Van Cleve's strengthened argument is surely sound. As an interpretation of part of Kant's theory of our attribution of necessity to propositions of arithmetic and geometry, it cannot be denied. It seems to me, however, the part of Kant's theory it actually pertains to consists in Kant's *disavowal* of any such attributions of necessity to the propositions in question. Kant surely could have reasoned things through as soundly as Russell and Van Cleve have reasoned them and therefore could have come to the same conclusion that they have reached. So we might ask what reasoning Kant could have gone through for him to think that his Copernican Revolution in philosophy succeeded in explaining our attribution of necessity to the propositions of

10 Van Cleve, ibid., p. 40.
11 Ibid.

arithmetic and geometry, as well as to those belonging to what he calls the general doctrine of motion.

[4] Defense of Kant. The modification of Van Cleve's strengthened argument I am about to offer emanates from a single change in his argument. It employs a logical relation that has already been introduced into our discussion, in chapter 5. I am going to substitute the relation of presupposition for that of entailment as the logical relation between a given proposition p and the ascription of the factor F to the proposition. By the term 'presupposition' I will be adopting Frege's and Strawson's use of the term.[12] According to that use, as explained earlier, one proposition presupposes another if and only if the truth of the latter is a logically necessary condition of *either* the truth *or* the falsehood of the former; that is, it is a logically necessary condition of the first proposition's having a truth value at all. The point can be made in terms of *entailment* itself: A proposition's merely having a truth value *entails* that any proposition it presupposes is true. So, according to the entailment, if such a presupposition is false, it is false that the proposition has a truth value.

I will symbolize the relation of presupposition by the triple-line arrow '$\Rrightarrow$'. The modified first premise now reads, that for any given proposition p, if p is necessary then p presupposes that p is a deliverance of our form of intuition.

1.* $(p)(\Box p \rightarrow [p \Rrightarrow Fp])$

Next, we add Van Cleve's Contingency Premise

2. $(p) \Diamond \sim Fp$

And, following Van Cleve, assume for a reductio

3. $\Box p$

Hence, from 3 and 1*,

4.* $p \Rrightarrow Fp$ (from 3 and 1*)

But now, instead of the modal theorem to the effect that if two propositions are related by entailment and the entailed proposition is possibly false so is the proposition that entails it, we get a modification of the theorem based on the notion of presupposition. It holds that if two propo-

12 For Strawson's more technical account of his use of the term than that provided in "On Referring," see his *Introduction to Logical Theory* (New York: John Wiley & Sons, Inc., 1952), pp. 175 ff.

sitions are related by presupposition and the presupposed proposition is possibly false, the presupposing proposition is possibly neither true nor false. If we symbolize a proposition's being either true or false as TVp, we get from 4*, 2, and the presuppositional modification of the modal principle

5.* $\diamond$ ~TVp.

Thus we get the proposition that p is possibly neither true nor false. This is to be contrasted with the penultimate conclusion of Van Cleve's strengthened argument (viz., proposition 5) that p is possibly false, which in turn entailed the conclusion that (necessarily p) is false.

And from 5* we get

6.* $\diamond$ ~TV$\Box$p

That is, we get the conclusion that (necessarily p) is possibly neither true nor false. Since 6* does not contradict 3 (as Van Cleve's 6 contradicts 3), Van Cleve's reductio does not go through.

This completes the defense of Kant against Van Cleve's strengthened argument, which was supposed to vindicate Russell's criticism of Kant. As such, however, the defense is only a negative argument in support of Kant. A positive argument on Kant's behalf will be provided in sections 6–8 below.

[5] A *Logical* Difference Between Initiators and Their Appearances. The difference between entailment and presupposition yields two different concepts of objects. The objects appearing as subject terms of the propositions that *entail* that the propositions are deliverances of our form of intuition are objects that are logically and causally independent of that form. The objects of a given arithmetic or geometric proposition in Van Cleve's strengthened argument have to be logically and causally independent of our form of intuition, if it is to be possibly *false* that the proposition is delivered by our form of intuition. Any given proposition referring to such an object could have a truth value even if the proposition were not delivered by our form of intuition, since it would refer to such an object independently of our form of intuition. That is why the proposition could be possibly false instead of being possibly without a truth value altogether. According to the interpretation of Kant that has been urged since the start of the book, those objects are *initiators.* In chapter 2 I defined them epistemologically as the external objects that get our knowledge started. As already noted in the Preface, this definition is to be compared with the logical one in my previous book,

where I took them to be the objects of transcendental affirmation. In a word, Van Cleve's use of entailment takes the object of the proposition – an initiator – *realistically.*

Exactly the opposite holds in the case of presupposition. Now the object that p refers to *would* depend on our form of intuition. However, in chapter 2, section 9, chapter 3, section 4, and chapter 4, section 2, it has been argued that for Kant a proposition referring to an object that is an *appearance* would depend on a form of intuition only if we were to intuit an object of the first sort, viz., an object that is logically and causally independent of the form, again, an initiator.[13] If the object that we would thus have to intuit were not logically and causally independent of the form, our intuition of it would cease to be a condition of the other object's (the appearance's) dependence on the form. Moreover, if we were not to intuit a logically and causally independent object, the dependent object (the appearance) would not exist. Even though a reference to such a dependent object would be possible, it would not "take place," as Kant puts it (A19/B 33, A33/B 49).[14] The possibility consists in the reference, but in this instance it would be a reference without any existing object being referred to. Consequently, as determined by the nature of presupposition, any given proposition belonging to arithmetic or geometry would not have a truth value. In sum, in contrast to Van Cleve's use of entailment, my use of presupposition takes the objects of the propositions in question *idealistically:* They are, as noted throughout, Kant's *appearances.*

The difference I am pointing out between Van Cleve's use of entailment and my use of presupposition and the resulting difference in different concepts of objects are the reasons I said earlier that Van Cleve's strengthened argument is surely sound as an interpretation of *part* of Kant's theory of the necessity of the propositions of geometry. The part in question is Kant's consideration of the reference of such propositions to objects that are independent of our form of intuition. Russell and

---

13 This entails that the concept of these logically independent objects – *initiators* – is not Kant's concept of things in themselves, since the former allows us to intuit them (and indeed is part of his theory that we do intuit them), whereas for Kant we cannot intuit things in themselves. For a concept of these objects that is expressed independently of the epistemological condition that they are constituents of the presupposed fact that we can intuit them (where *intuition* is an epistemological concept), but rather in terms of Kant's transcendental logic, see the author's *Kant's Theory,* chaps. 1 and 2.

14 For an earlier mention of an intuition's "taking place" see chapter 2, section 9.

Van Cleve are right: Our form of intuition cannot explain any legitimate attribution of necessity to such propositions that refer to such independent objects. In other words, Kant's account of *de re* necessity in the matter of the necessary truths of geometry themselves being necessarily true of objects would be devastated.

But this is the part of Kant's theory in which he *disavows* such attribution. As I have already noted, he could have reasoned through to the same conclusion reached by Russell and Van Cleve as surely as they could reason through to it. That is why, I submit, he took his Copernican Revolution in philosophy to entail his transcendental idealism. It is only appearances, and not objects that are logically and causally independent of our form of intuition, that are the objects of the necessary truths of geometry.

[6] Preliminary to the Derivation of the *De Re* Necessity of Geometry. In contrast to the negative argument given above in support of Kant's theory of our attribution of necessity to the propositions of arithmetic and geometry, and hence the possibility of the *de re* necessity of ascribing necessary mathematical properties and relations to appearances – the argument that was given as a rebuttal of Van Cleve's strengthened argument, I still owe the reader a positive argument that will support Kant's use of *de re* necessity, at least in regard to the ascription of the necessary properties and relations or necessary truths of at least geometry to appearances.

However, before I proceed with the derivation, I must limit the claim of the argument about to be given. The connection for Kant between the truths of arithmetic, and for that matter, the truths of the general doctrine of motion, on the one hand, and time, on the other, is a complicated matter. It is much less straight forward than the connection for Kant between the truths of Euclidean geometry and Euclidean space of which I am about to speak. The same limitation applies to the necessary truths Kant claims to find among the principles of synthetic judgments.[15] So, I am going to limit the rest of my argument in this part of the chapter to the connection I take Kant to find between the necessity we ascribe to the truths of Euclidean geometry and Euclidean space.

[7] Derivation of the *De Re* Necessity of Euclidean Geometry. I have used the difference between presupposition and entailment to distinguish appearances from objects that are independent of our form of intuition (i. e. initiators) on the ground that the ascription of our form of intuition

15 The limitation will be waived in the next part, on causality.

to appearances *presupposes* that we intuit objects that are logically and causally independent of our intuition of them, i.e. the initiators, and hence are logically and causally independent of our form of intuition. Our ascription of our form of intuition to the initiators, on the other hand, instead *entails* that we intuit them. We thus use presupposition to begin the derivation of Kant's use of *de re* necessity with regard to Euclidean geometric properties and relations, or the truths that express them, that are ascribed to appearances.

Let S_1 be the proposition that outer appearances have Euclidean-spatial properties and relations.

(1) S_1 presupposes the contingent proposition that we intuit initiators as "outside us" (A22/B37).

That is, (2) if S_1 has a truth value, we in fact intuit initiators as outside us.

Furthermore, (3) it is a conceptual truth for Kant that if we intuit initiators as outside us, we do so through our form of outer intuition.

Moreover, (4) our form of outer intuition is Euclidean-spatial in nature.

Therefore, (5) if we intuit initiators as outside us, we do so through our Euclidean-spatial form of outer intuition.

Therefore, (6) if we intuit initiators as outside us, the objects of our intuitions – outer appearances of initiators – have Euclidean-spatial properties and relations.

From (1) through (6) we can conclude

(7) S_1 *has a truth value only if it is true.*

If (7) were contingent, S_1 might be false, since if either (4), (5), or (6) were false, (7) would still follow from (1)-(6) – it would simply be a case of a contingent proposition following from a set of premises, some of which would themselves be contingent, and hence possibly false. We could not then conclude that S_1 *could not* be false. And if S_1 might be false, it would not be necessary. So, if this argument, (1)-(7), is going to demonstrate the necessity of S_1, (7) must be necessary. But it cannot be necessary unless (1)-(6) are necessary.

As a true proposition about a logical relation between two propositions, (1) is necessary. (2) is necessary because it is an explication of (1). Being a conceptual truth for Kant, (3) is necessary. The necessity of (4) follows from an argument in section 4 of the previous chapter. There it was argued that Kant identifies outer intuition with space and inner intuition with time through his attempt to establish the connections of space and time with outer and inner sense, respectively – an attempt that precedes the Metaphysical Expositions of the concepts of space and time (A22–23/B37). If, then, (4) is necessary, so is (5), and hence, so is (6). Consequently, (1)-(6) are necessary, and therefore, (7) is necessary as well.

Therefore, (8) if S_1 has a truth value, it *cannot* be false.

For if *per impossible* it were false, it *would* have a truth value, i. e. false, yet the contrapositive of (7) states that if S_1 were not true, it would *not* have a truth value. In a word, since (7) is necessary, it is *impossible* that *both* S_1 has a truth value *and* it is false. (This reasoning, called 'R_1', will be cited as the justification of premise 8 in the Derivation of Causality in section 16 [Part II] below, *mutatis mutandis.*) Since it is *impossible* for S_1 to be false,

(9) if S_1 has a truth value, S_1 *must* be true.

At this point it may be objected that my argument for the necessary truth of S_1 is overburdened by what for some may be a dubious doctrine from logical theory, viz., the theory of presupposition and all its trappings about propositions that have sense but no truth value. As explained in chapter 5, Russell's semantics is much simpler: propositions without truth values are meaningless. Fregean sense without reference, or meaning, is an unnecessary complication in our understanding of how terms refer to objects.

The objection is supported by the logical fact that the *a priori* character of S_1, and hence its necessity – the necessity that is logically connected to the *a priori* – follows from (3)-(6) alone – that is, *independently* of (1) and (2). For since (3)-(6) are admittedly *a priori*, S_1 must be *a priori* as well. So, why complicate the argument by introducing the logical relation of presupposition into it? Kant interpretation has gotten on quite well so far by relying on just (3)-(6) to account for the necessity of S_1. Since our interest is in Kant's argument for the *a priori* necessity of S_1, there is no need to bring into the argument possibly controversial complications of presupposition from the field of logical theory.

One of the main arguments of the book, however, is that Kant's uses of *de re* necessity must be both logically and epistemologically related to his use of the concept of existence. Consequently, our *de re a priori* propositions must be related to our intuitions of initiators, if our claims of necessary truth are to be based on what exists. And it is not enough to meet the demands Kant puts on such claims to argue that the relation to existence Kant has in mind is fully satisfied by considering our intuitions of initiators as a *sufficient* condition of the ascription of Euclidean-spatial properties and relations to objects. But this is all that premises (3)-(6) accomplish. On this view of Kant's intentions, whereas Euclidean-spatial properties and relations are *necessary* conditions of our intuitions of initiators, those same intuitions are not *themselves necessary* conditions of the ascriptions of those properties and relations to initiators – as *appearances*, to be sure.

The position the book has taken, however, is that this view of Kant's theory is mistaken. Besides the book's emphasis throughout that the *a priori* depends in certain ways on existence, chapter 4 makes the dependence in the present connection fully explicit: Our empirical intuition of initiators is a *necessary* condition of our ascription of existence in space and time to them. This is a major issue between Guyer and me.

Without propositions (1) and (2), the argument for the necessity of S_1would leave our intuitions of initiators, and thus their existence, as merely *sufficient* conditions of the necessity, as they are in (3)-(6). The inclusion of (1) and (2) as ineliminable premises in our argument for the necessity of S_1 make our intuitions of initiators *necessary.* That is the reply to the objection under consideration.

The application of (9) to Euclidean geometry is as follows.

(10) S_1 could not be the necessary truth I have just argued it is, if Euclidean geometry – the geometry of those properties and relations in relation to appearances – were not necessary.

Finally, therefore, (11) if S_1 has a truth value, the propositions of geometry are necessary truths of outer appearances.

[8] Observations on the Derivation. The derivation needs several remarks, if its claims are to be understood as intended. The first concerns step 4. Kant maintains that it is *a priori*, and I have so argued that it is in his defense. Can it be contingent as well? If 'contingent' is used here in the sense that it is 'metaphysically possible' that the form of our outer intuition might have been non-Euclidean-spatial, in Kripke's sense of 'metaphysical possibility,' Kant's claim could still be *a priori*, and hence necessary in the epistemic sense that Euclidean space determines our outer intuition, and hence that Euclid's geometry is necessary with respect to our knowledge of objects. This would then be a case of Kripke's *epistemic necessity.* If so, the proposition would be true only in regard to what we can know about objects, but not true in all possible worlds, regardless of what we can know, if anything, about the objects in those worlds.

So, at B145, where Kant asserts the futility of trying to explain the proposition in question, despite the fact that it is *a priori* – something that does not interfere with Kant's explanations of other *a priori* propositions – two things follow. First, the proposition for Kant is not analytic. For if it were, an explanation would have been available to him. It could have been couched in terms of a subject-concept's containing a predicate-concept. The fact that Kant thinks that an attempt at such an explanation would be futile, indicates that the proposition not analytic.

The same consideration also indicates that the proposition is only epistemically, but not metaphysically, necessary. For if it were metaphysically necessary, there would seem to be another explanation available to Kant, that is, one besides its being analytic, if he had thought it were so. It would be something to the effect that the terms 'the form of our outer intuition' and 'Euclidean space' are Kripkean *rigid designators* and that the proposition expressing an identity between them is true. Kant could therefore claim that since the two terms are rigid designators and the identity is true, it is *metaphysically impossible* for the form of our outer intuition to be anything but Euclidean space. In other words, the explanation would be that the truth of the identity – that the form *is* Euclidean space – is metaphysically necessary. We would have a case in which metaphysical and epistemic necessity coincide.

It would then seem odd, however, that Kant resists any attempt to explain the identity, since the identity for him expresses just a "peculiarity" (*Eigentümlichkeit*) of our outer intuition, viz., that its form is Euclidean space. For example, he does not say that it is just a "peculiarity" of our outer appearances that they are Euclidean spatial. If he had, he could not have legitimately offered the explanation that Euclidean geometry must refer to our outer appearances *because* of our outer sense (B41). Since it is *a priori* that appearances are subject to the condition of our outer sense, and since it is also *a priori* that the form of our outer sense is the form of our outer intuition, and since, finally, it is *a priori* that the latter is Euclidean space, it *follows* that it is *a priori* that Euclidean geometry refers to outer appearances. In other words, Euclidean geometry must refer to outer appearances. That, in brief, is Kant's *Transcendental* Exposition of the concept of space.

A further observation is that this *necessary* reference of Euclidean geometry to our outer appearances, which is due to our outer sense, is connected to step (2) of the derivation. First, we deal with the necessity of the reference and take note that it is due to outer sense: If and only if appearances are represented by means of outer sense (as outside us) can Euclidean geometry refer to them. On the other hand, the ascription of Euclidean-spatial necessary properties and relations to outer appearances, i.e.

S_1, consists in the reference of Euclidean geometry to the appearances. So, if S_1 has a truth value, Euclidean geometry has that reference. Finally, if S_1 is a *consequence* of the representation of the appearances by means of outer sense, S_1 is said to be *necessary* in that sense. But that is the very sense in which the reference of Euclidean geometry to outer appearances is said to be necessary, viz., that outer appearances are represented by means of outer sense.

[9] Symbolization of the Derivation. We can symbolize the argument in support of Kant's theory of the *de re* necessity with which the necessary truths of geometry are ascribed to appearances. The numbered symbols in the following argument correspond to the parenthetically numbered propositions of the argument that has just been given.

Before proceeding, it should be acknowledged that what follows is merely a symbolization of the preceding argument (numbered (1) through (11)) whose validity depends on the meaning or content of the particular propositions that are represented, i.e. S_1, S_2, etc. Consequently, its validity cannot be determined simply on the basis of the logical relations (syntax) of the symbols S_1, S_2, etc. That is, it is not a formal proof of its conclusion. Chapter 10 will present more formal arguments for some of my later assertions – arguments that are less dependent on the content of the propositions involved, but even these will depend on certain constraints that I adopt. Consequently, it will not be claimed with respect to them, either, that they are strictly formal proofs.

As before, let S_1 stand for the proposition that outer appearances are Euclidean-spatial

let S_2 stand for the contingent proposition that we intuit initiators as outside us (A 22/B37),

let S_3 stand for the conceptual truth for Kant that we intuit objects only through our form of intuition

let S_4 stand for the *a priori* proposition that our form of outer intuition is Euclidean-spatial,

let S_5 stand for the *a priori* proposition that we intuit initiators as outside us through our Euclidean-spatial form,

and let G stand for the propositions of Euclidean geometry.

Again let the single-line arrow, '$\rightarrow$', stand for material implication, let the double-line arrow, '$\Rightarrow$', stand for entailment, let the triple-line arrow, '$\Rrightarrow$' stand for presupposition, let '$\Box$' stand for necessity, let '$\Diamond$' stand for possibility, and let 'TV' stand for having a truth value.

1. $S_1 \Rrightarrow S_2$ (A19/B 33, A33/B 49, and A22/B 37)

(i. e. S_1 presupposes that we intuit initiators and we intuit them as outside us.) (This proposition symbolizes the proposition (1) that was given above in the more informal proof.)

2. $TV_{S1} \Rightarrow S_2$ (from 1 and the meaning of "presupposition")

3. $S_2 \Rightarrow S_3$ (a conceptual, and thus *a priori* truth for Kant)

4. S_4 (an *a priori* truth about our outer intuition)

5. $S_2 \rightarrow S_5$ (from 3 and 4)

6. $S_2 \rightarrow S_1$ (from 1, 5, and Kant's theory that if we intuit initiators, the objects of our intuitions [i.e. appearances] have the form of the intuitions of which they are appearances)

7. $TV_{S1} \rightarrow S_1$ (from 2 and 6)

8. $\Box(TV_{S1} \rightarrow S_1)$ (from the *a priority* of (1–6) and transitivity)

9. $TV_{S1} \rightarrow \sim\Diamond\sim S_1$ (from 7 by contraposition and the impossibility of both $\sim TV_{S1}$ and $\sim S_1$)

10. $TV_{S1} \rightarrow \Box S_1$ (from 9)

11. $\Box S_1 \Rightarrow \Box G$ (Euclidean geometry is the geometry of Euclidean space)

12. $TV_{S1} \rightarrow \Box G$ (from 10 and 11)

This completes the derivation of *de re* necessary truths from certain *a priori* truths plus the logical relation of presupposition, which is necessary to capture Kant's intention to demonstrate *a priori* the reference of necessary propositions to existence. The success of the proof depends on drawing the distinction between entailment and presupposition, and thus depends on drawing the distinction between initiators and appearances as the respective objects of the two types of ascription of necessary spatial properties and relations: one where the ascription entails that we intuit initiators, and the other where the ascription presupposes that we intuit them. It thus depends on drawing the implication of Kant's transcendental idealism from his Copernican Revolution in philosophy, where the transcendental distinction between appearances and initiators is now based on the logical difference between presupposition and entailment as well as the epistemological requirement that existence must be given through our empirical intuitions of initiators.[16]

[10] Recapitulation. In this chapter I have first argued that Kant's transcendental idealism is crucial to his theory of the possibility of *a priori* knowledge. In support of this thesis, I have shown how the idealism defends Kant from the objections of Russell and Van Cleve. They argue that the contingency that our mind is constituted in the way that it is makes it impossible for the truths of geometry to be *de re* necessary, according to Kant's theory that our cognitive constitution explains the possibility of the necessity. The (metaphysical) contingency of the actual constitution of our mind destroys the *de re* necessity that is supposed to depend on that constitution. Hence, Kant cannot explain the *de re* necessity of geometry in terms of our cognitive constitution.

Transcendental idealism has been essential to my defense of Kant. I argued that the logical relation of a proposition of geometry to the ascription of a form of our intuition to the proposition is presupposition, not entailment, the relation employed by Van Cleve. My analysis was intended to show that whereas presupposition takes the objects of geometry as *appearances*, entailment takes them as things in themselves.[17] Since Kant's theory of the necessity of geometric truths requires that the objects are appearances and not things in themselves, my use of presupposition accords transcendental idealism the central role that Kant gives it in his explanation of the necessity.

[11] Existence Again. In chapter 3, I took the opposite tack and argued that the criterion of existence belonging to Kant's theory of knowledge is independent of his transcendental idealism. The criterion determines that objects can be said to exist just in case they are objects that can be given through intuition. Since both (a) objects that are independent of sensibility (i. e. initiators) can be given to us as appearances through our sensible intuition and (b) objects can be given to God as they are in themselves through intellectual intuition, the objects, whether in (a) or in

16 At this point, I would like to thank Robert Hanna for his helpful review of material in this chapter, as well as material in chapter 3, and for his suggestions for its improvement. Of course, any deficiencies in the chapters remain my own responsibility. He finds that the interpretation of the *Critique* in this chapter is tantamount to his own model theoretic interpretation of the *Critique*. See his *Kant and the Foundations of Analytic Philosophy* (Oxford: Clarendon/Oxford University Press, 2001), esp. pp. 239–264.

17 And not merely *initiators,* since we are now considering initiators *as determined*, by Euclidean space.

(b), can be said to exist according to the same criterion of existence. But since the criterion determines the ascription of existence to objects, whether they are initiators, appearances, or things in themselves, it is independent of the transcendental distinctions between them. However, because the criterion is both unique and general, and yet, as already noted in chapter 3, hardly recognized among Kant scholars, it can inadvertently drive the continuing controversy over the correct interpretation of the idealism. To repeat, its uniqueness can encourage the interpretation that transcendental idealism consists of a single set of objects say, appearances, bearing dual aspects. On the other hand, its generality might encourage scholars to hold that its coverage can be divided between two sets of objects – appearances and things in themselves. Having lent its tacit support to both sides in the dispute over the correct interpretation of the dualism belonging to the idealism, the criterion may in fact be contributing to the continuation of the controversy. Perhaps awareness of this possibility, I conjectured, may take some of the intensity out of the dispute and even lead it in a direction in which it could become more manageable, if not resolvable.[18]

[12] Existence and *De Re* Necessity. Chapter 3 and this chapter, and their two central ideas, existence and *de re* necessity, can be summarily connected here. The existence of initiators that figure in the presuppositions of the necessary propositions of geometry, for example, is determined by the same criterion of existence that is independent of the very idealism that explains the *de re* necessity with which those propositions are true of appearances. That is precisely why the criterion of existence in general can underlie Kant's account of the *de re* necessity of geometry with respect to appearances. Therefore, factors that appear to stand in opposition to each other – existence and *de re* necessity – actually complement each other.

II Causality

[13] Preliminaries. Just as we applied our interpretation of Kant's uses of *de re* necessity to only one of the two *a priori* concepts analyzed in the Aesthetic, viz., the concept of space, so we will apply the same interpre-

---

18 For an initial discussion of the possible role of the criterion of existence in the controversy over the correct interpretation of transcendental idealism see chapter 3, section 7.

tation to just one of his categories, viz., causality, whose necessary determination of a manifold of both a sensible intuition in general[19] and a temporal intuition in particular[20] (henceforth abbreviated simply as "necessary determination of a manifold of an intuition") is, alongside the necessary determination of a manifold belonging to the other categories, the *equivalence* of the objective of the Transcendental Deduction. Thus, the interpretation ought to be applicable, *mutatis mutandis,* to the other categories as well. But first I must say a word or two about the limited nature of the treatment causality is going to receive in the discussion that follows.

First, the issue at hand is said to be only the *equivalence* of the objective of the Transcendental Deduction, not the objective itself. The latter is obviously the demonstration of the necessary reference (*Beziehung*) and application (*Anwendung*) of the categories to appearances. It might be the case, Kant avers, that appearances are not subject to the conditions of the unity of the understanding (A90/B123). The job of the Deduction is to bypass that possibility.

Appearances, however, are particular objects of a manifold of both a sensible intuition in general and a temporal intuition in particular. Consequently, if it can be shown that the categories must determine the manifolds of both, it must determine appearances as well. On the other hand, short of that demonstration, Kant will not have successfully bypassed the otherwise unavoidable possibility that appearances might not be subject to the categories. Therefore, Kant will be able to show that the categories must refer and apply to appearances if and only if he can show that the categories must determine both manifolds – that of a sensible intuition in general and that of a temporal intuition in particular. That is why the most direct route to an understanding of Kant's strategy in the Transcendental Deduction goes through his demonstration of what has just been called the *equivalence* of the objective of the Deduction.

And second, our treatment of causality will not deal with Kant's *schematic* implementation of causality with time, and thus will leave untouched many of the salient features of the principle of objective succession as the principle appears in the Second Analogy of Experience. Nonetheless, it will still employ the idea of objective succession for its own purposes. In the Transcendental Deduction, especially the B-Deduction, objectivity is the major consequence of Kant's claim that the categories must

19 What Dieter Henrich calls "step one" of the B-Transcendental Deduction.
20 What Dieter Henrich calls "step two" of the same Deduction.

determine the manifold of a sensible intuition in general (B143), and succession is part of his claim that a temporal succession of a manifold of an intuition must be subject to the categories, if the succession is to be objective (B161), as illustrated in the succession of states involved in the alteration of some water from its liquid to its solid state (B162–3).

[14] Two Concepts of a Succession of a Manifold of an Intuition. In the discussion of space in section 5 of this chapter, the logical difference between presupposition and entailment was used to distinguish between two concepts of Kantian objects, appearances and things in themselves, both of which devolved onto the concept of an initiator. Since the distinction between presupposition and entailment is going to be crucial to the immediately forthcoming analysis of Kant's treatment of the concept of causality in its occurrence in the Transcendental Deduction, we will find a corresponding distinction between two concepts of a succession of a manifold of an intuition: *as thought by me* and *as external to my thought*. It will be presupposition alone that provides for the former, about which it is going to be said that it must be thought through the concept of causality, if the succession is to be objective (B161). Entailment, and thus such a succession as external to my thought, will not do the job.[21]

[15] The Argument for Presupposition. As already noted in chapter 8, section 5, the problem the Transcendental Deduction is supposed to solve originates in the logical independence between sensibility and the understanding, and therefore between a manifold of an intuition of sensibility and the categories. The independence entails that the manifold might fail to be subject to the categories, or as we have equivalently framed the problem in section 13 of this chapter, that the categories might fail to apply to appearances, and thus might be empty of objects that can be given to us through intuition (A90/B123).

The problem of the Transcendental Deduction not only originates in the logical independence between sensibility and the understanding, however; it also cannot be resolved if the independence is not somehow bypassed. A manifold of an intuition of sensibility must somehow be relativized to my thought, if it is to be necessarily subject to the categories, and for our present concern, subject to causality. Given that concern, it should be understood that our discussion of the categories that follows will therefore be limited to our treatment of causality. We should then say that the independence between sensibility and the understanding

21 This reasoning tracks that followed in section 5 of chapter 6.

will make it impossible to demonstrate through a transcendental deduction that causality must determine a succession of a manifold of an intuition of sensibility: Any justification (*quid juris*) (A84/B117) of the determination of a succession of such a manifold will be out of Kant's reach.

For Kant, the independence persists, however, so long as causality is supposed to determine such a succession considered in the *disabling* sense of its being external to my thought: Unless such a succession is considered as conditioned by, or relative to, my thought, it cannot be *necessarily* determined by causality. It is the mediating role of *my thought* that solves the problem of the deduction of causality for Kant. This is similar to the role we have already seen (in section 4 of the previous chapter) played by outer and inner sense with respect to the *necessary* spatial and temporal determinations of initiators: These two *a priori* senses are the *means by which* initiators must be represented, if it is to be necessary that they exist in space and/or time, as outer and inner appearances, respectively. So my thought is the necessary *means by which* such a succession must be represented, if it is to be necessary that it is subject to causality.

The condition of my thinking that enables causality to necessarily determine such a succession Kant calls *the transcendental unity of apperception* – the 'I think' – or what he calls, in §19, in the Transcendental Deduction, "objective consciousness."[22] Apart from being represented by means of the 'I think', every manifold of a sensible intuition, and thus every succession of such a manifold, might fail to occupy a place in a causal nexus. On the other hand, once such a succession is represented by means of the 'I think', it must occupy such a place. The necessary causation follows from the for Kant *a priori* knowledge that objective consciousness depends on judgment (B141–2), which in turn depends on the employment of the logical function of a *hypothetical* judgment, if the succession of a manifold of an intuition in general is subject to the 'I think', and this is so if and only if the succession is thought through causality (B128, B143). This very reasoning, *R*, which will be cited as 'R_2' in the justification of premise 4 in the Derivation of Causality that will be given in the next section of this chapter, leads to the conclusion that an appearance must be subject to causality if and only if, as an instance of an objective succession of a manifold of a sensible intuition in general, it must be subject to the 'I think', and being thus subject, de-

22 For a fuller discussion of the role of the 'I think' in enabling the categories to be necessarily ascribed to appearances in space and time see chapter 6, section 1 *et passim*.

pends on its being categorized through causality. Consequently, the problem of the deduction of causality remains unresolved for Kant, if the necessary role of the 'I think' in determining the succession of a manifold of a sensible intuition (and its dependence on causality through the logical function of a hypothetical judgment) is omitted from the deduction. In such a case, a succession of appearances (as a temporal succession of mere representations) would be "nothing to me" (B132); that is, it would not be an *objective* succession. In that case, a succession of appearances (as a succession of mere representations) would remain *external* to my thought. What, then, is the logical relation between a proposition about a succession of a manifold of a sensible intuition, whether in general or a temporal intuition in particular, and its being subject to the 'I think'? Is it entailment or presupposition? The answer to this question will, given the equivalence described in sections 13 and 15 above, determine the answer to the same question about the logical relation between a proposition about a succession of appearances as a succession of representations and its being subject to the 'I think'.

The difference between entailment and presupposition can explain the presence or the absence of Kant's solution to the problem of the deduction of causality. It cannot be the case that causality *must* determine both manifolds in question, i.e. that of a sensible intuition in general and that of a temporal intuition, if that very proposition, i.e. that causality must determine them, *entails* that either manifold must be subject to the 'I think', or more simply, what will be adequate for our purposes, entails that the manifold of a sensible intuition in general alone must be subject to the 'I think'. For if the relation were entailment, and though Kant claims to know *a priori* that causality is indeed a form of our thought of objects of sensible intuition in general, he is as open about his lack of any *proof* that a form of such thought must be causality as he is about the lack of any proof that space and time must be our forms of intuition, even though he also claims to know *a priori* that they are our forms of intuition (B144–5). So, even though he says that he knows *a priori* that causality is a form of such thought, that it *must* be a form of such thought is, he also says, beyond his capacity to demonstrate. In other words, if one insists that it *must* be a form of such thought, the necessity invoked is only *epistemic*, not metaphysical.[23] For if it were metaphysical, a reason that was independent of knowlege

23 For a brief discussion of the contrast between the two senses in which a proposition may be necessary, see the next chapter, section 1 *et passim*.

would be forthcoming, perhaps in the form of a principle with implications for possible objects or states of affairs that would be independent of knowledge. It follows that I might be able to think a manifold of a sensible intuition in general without causality, which indeed is a complaint that has actually been made against Kant's claim that causality is one of the necessary forms of our thought of objects of sensible intuition in general. All this remains the case even though Kant may be right that he knows *a priori* that causality indeed is one of the ways in which I think such objects. This would entail that while Kant may be right that he knows *a priori* that the logical function of a hypothetical judgment is one way in which I can bring "various representations under one common representation" (A68B93), and thus in that act bring a manifold of a sensible intuition under a common representation, it still *might* be the case that I could bring every such manifold under a common representation without causality.

Indeed, at least from Frege on it has been argued that Kant was unnecessarily circumscribed by the teachings of the logic that was traditional at his time. Though this is definitely not to equate Kant's logical *function* of a hypothetical judgment, which belongs to what he calls *transcendental logic*, with a logical *form* of judgment, which belongs to what he calls *general logic*,[24] it is an acknowledgement that there is a logical relation between the two logics, the latter being more general than the former. Given that relation, the criticism of Kant's insistence on the inescapability of causality often originates in a criticism of what have come to be considered as the unnecessarily narrow confines of his general logic. For it has been demonstrated in modern logic that the logical form of a hypothetical proposition can be reduced to a more general form of proposition that employs only the logical operations of conjunction and negation. This is not to mention the criticism of his claim for the necessity of causality which originates in physics.

The criticism of Kant that is based on logic can, however, on Kant's behalf, be said to be unfounded. The greater generality of general logic allows it to dispense with certain logical *forms* of judgment that might correlate with certain logical *functions* of judgment that are required specifically for relations among cognitions that are exclusively *a priori*. For the claims of general logic cover all knowledge, regardless of its origins, *a priori* or empirical (A55/B80). Therefore, that general logic can do

24 See chapter 9, *Kant's Theory,* where I argue that Kant's logical functions of judgment are not to be equated with his logical forms of judgment.

without the hypothetical *form* of judgment is no argument against Kant's claim that *transcendental* logic needs the logical *function* of judgment, since the latter, but not the former, provides the logical basis of causality – a concept whose deduction belongs to transcendental philosophy alone.

Given this lack of a metaphysical necessity pertaining to my employment of causality, if the proposition that causality must refer and apply to appearances were to *entail* that I must think a manifold of a sensible intuition in general, then, in a situation in which I were able to think such a manifold *without* causality, it would be *false* that causality must determine a succession of a manifold of a sensible intuition in general; and that falsehood would mean the failure of the deduction of causality.

The falsehood of the equivalent of the very proposition that the deduction is concerned to prove is obviated, however, if presupposition is substituted for entailment as the logical relation between the proposition in question, i. e. that causality must determine a succession of a manifold of a sensible intuition in general, and the proposition that I must think a manifold of a sensible intuition in general (if the manifold is to be combined in an object). For then, since it is presupposed that I think a manifold of a sensible intuition in general, and further, since it is the case, and Kant claims to know *a priori* that it is the case, that I would not have such a thought unless some of my thoughts were causal, then, in case *none* of my thoughts *were* causal, the presupposition would *not* be satisfied, and hence the proposition in question (that causality must determine a succession of a manifold of a sensible intuition in general), instead of being simply false, would not have a truth value. In other words, the very proposition that causality must determine a succession of a manifold of a sensible intuition in general would *itself* cease to refer to such a succession, i. e. to an *objective succession* – one in which the manifold is *combined* in an object, its truth value having been vacated by the failure of the presupposition. The question of the determination by causality of a succession of a manifold of a sensible intuition as *objective* would then not even arise. It would thus cease to be a question that needed to be addressed by a transcendental deduction.

[16] Derivation of the *De Re* Necessity of Causality. Again let the single-line arrow, '$\rightarrow$', stand for material implication, let the double-line arrow, '$\Rightarrow$', stand for entailment, let the triple-line arrow, '$\Rrightarrow$' stand for presupposition, let '$\Box$' stand for necessity, let '$\Diamond$' stand for possibility, and let 'TV' stand for having a truth value.

Let S_1, S_2, S_3, and S_4 stand for the following four propositions, respectively:

S_1: every objective succession of occurrences must be thought causally.

S_2: I think a succession of a manifold of a sensible intuition as objective.

S_3: I think the *form* of thinking a succession of a manifold of a sensible intuition as objective.

S_4: I think causally

1. $S_1 \Rrightarrow S_2$ (a conceptual, *a priori* truth)
2. $TV_{S1} \Rightarrow S_2$ (from 1)
3. $S_2 \Rightarrow S_3$ (a conceptual, *a priori* truth)
4. The *form* in S_3 is causality (reasoning R_2 in section 15)
5. $S_2 \rightarrow S_4$ (from 3 and 4)
6. $TV_{S1} \rightarrow S_1$ is true (from 2 and 5)
7. $\Box TV_{S1} \rightarrow S_1$ is true (1–4 are *a priori*; thus, so are 5 and 6)
8. $TV_{S1} \rightarrow \sim\Diamond\sim S_1$ (reasoning R_1 in section 7)
9. $TV_{S1} \rightarrow \Box S_1$ (from 8)

With this derivation of causality, we have completed the defense of Kant's theory of *de re* necessity from challenges that originate in the possibility that our cognitive constitution might have been different from what it is – so-called challenges from "within." When combined with our defenses of Kant against Quine's challenges from "without" – those pertaining to the very intelligibility of the ascription of putative necessary properties and relations to objects that are referred to independently of the properties and relations – we have covered prominent, serious attacks on Kant from opposite directions which were mounted in the last century by analytic philosophers, whether Kant was the intended recipient of the attacks or not. Finally, to a large extent, the defense of Kant was carried out in terms and techniques that were found within analytic philosophy itself.

Chapter 10 – Presupposition and Real Necessity

[1] Real Necessity and Metaphysical Necessity. Saul Kripke would pronounce the results of the last chapter (i. e. the necessity of Euclidean geometry and the principle of causality) examples of *epistemic*, not *metaphysical*, necessity. His treatment of necessity relegates the sense of the former necessity to a less than "highest degree" or "strictest possible sense" of the term, a sense that he accords metaphysical necessity alone.[1] This is the sense in which the necessity covers all possible worlds in which given objects exist independently of possible worlds in which the conditions of our knowledge exist, or obtain.[2]

Kant's does indeed limit the necessary truths to the possibility of our knowledge of the objects that they refer to. The question that arises for us in this context is that if Kant can limit the possibilities of objects according to certain *mental* facts about us, albeit facts that are claimed by him to be known by us *a priori*, why cannot the possibilities of objects be limited by certain particular facts about the objects themselves, facts that are independent of the mind and that may not exist in every conceivable or describable situation or state of affairs? As independent of the mind, the facts are characterized as *real* and their characterization as *limited* is meant to fit the requirement that the possible worlds in which they exist may be only some, but not necessarily all, worlds that are conceivable, describable, or otherwise "stipulated," as Kripke would say.[3] Henceforth, however, for the sake of convenience, the necessity whose idea I will be defending will be called simply *real necessity.* (The use of the term 'real' in respect to the modalities is an ironic adaptation of Kant's use of *reale* in the *Critique* at $Bxxvi_a$.)

---

1 In the following I am going to develop the interpretation of *de re* necessity in a notation that uses *single* quotation marks for mentioning symbols and technical terms, following the use of such marks by Carnap, Prior, Lemmon, Strawson, and Kripke. This is the same method I followed in chapter 6, where it was especially important that I do so, since I was there actually reproducing passages from Kaplan's "Quantifying In." See chapter 6, endnote 1.

2 Saul A. Kripke, *Naming and Necessity* (Cambridge, MA: Harvard University Press, 1980), pp. 99 and 125, respectively.

3 *Op. cit.* , p. 44.

Moreover, since these would be facts about objects whose existence and properties and relations are independent of the mind, it may be that they can be known only *a posteriori*. Of course, whether these facts can be, or are, known at all doesn't either make or prevent them from being facts about these objects, since the objects and their independent properties and relations are understood as being independent of the mind, and therefore as being independent of our capacity to know the facts about the objects. If the possibilities of objects can be so limited by such facts, we could be said to be following Kant's account of epistemic *de re* necessity with respect to the very idea that the possibilities of objects can be limited by facts, only in the present instance the facts that would determine these non-Kantian possibilities would be facts about independent, existing objects instead of facts about the mind.

While such an interpretation cannot be part of a theory of the possibility of our *a priori* knowledge of objects, it can be an interpretation of our uses of the necessity that, on the one hand, follows Kant's method in arriving at his account of epistemic necessity but, on the other hand, is determined by just certain facts that exist in reality instead of being determined by just certain facts about the mind. This interpretation of the necessity would at least have the virtue of seeming to satisfy a natural inclination of ours to be realistic about our application of necessity to propositions about objects as a *de re* necessity, while at the same time acknowledging its guiding principle to be a generalization from our interpretation of Kant's account of epistemic *de re* necessity, and hence acknowledging a debt to Kant, or at least to our interpretation of Kant.

The realistic interpretation of *de re* necessity will look like an interpretation of necessity that is often called 'physical necessity'; yet the former is really distinct from the latter. The difference is that since physical necessity is determined by physical laws, and such laws already contain the very necessity that stands in need of interpretation. That is, physical laws cannot account for physical necessity without circularity. On the other hand, the facts that are to determine the real necessity to be developed here will not involve us in such circular reasoning – they will not contain the very necessity whose interpretation is in question.

The realistic interpretation of the necessity will also be silent about whether such facts on which it depends accord with epistemic necessities. Again, these latter are determined according to mental facts, whether these facts are logically connected to our *a priori* knowledge of objects or can be known wholly *a posteriori*. By contrast, the real necessities to be interpreted here may *vary* from one set of mental facts to another.

Moreover, because the facts that determine these necessities are not mental, they are closer to Kripke's *metaphysical* necessities than they are to Kant's epistemic ones. Still, they are nonetheless explained independently of Kripke's semantics as well. Therefore, taking both Kant and Kripke together, these real necessities are independent of both of their interpretations.

There is a further difference between real necessities and Kripke's metaphysical necessities. In keeping with the limited aim announced in chapter 6, section 1 of interpreting Kant as employing a merely a logical or syntactical sense of *de re* necessity along the lines suggested by Kit Fine's discussion of Kaplan rather than the metaphysical sense of it associated with Kripke,[4] arguments for real necessities are devoid of the modal or counterfactual intuitive considerations that are basic to Kripke's arguments for metaphysical necessities that appear in *Naming and Necessity.* They are therefore also independent of any consideration of the essentialism that figures in certain of Kripke's uses of metaphysical necessity.

[2] Real Necessity Determined *A Priori.* The independence of real necessity from essentialism can be reached from another direction – from the top down instead of from the bottom up, to paraphrase Kant (A119). The propositions that correspond to the facts on which real necessity depends constitute a class of propositions that is determined by a higher level proposition. It is the trivial, but no less analytic, truth that the referent of a proposition does not exist unless it satisfies the necessary conditions of its existence, that is, those conditions that it satisfies, if it exists. As a proposition that collects the lower level propositions that correspond to the facts on which real necessity depends, this higher level proposition is fundamental to our interpretation of real necessity. Consequently, as both *a priori* and fundamental to our interpretation, this higher level proposition will be considered the *principle of real necessity.* It is a principle that excludes propositions to the effect that something is a condition of the existence of objects whose existence is presupposed by given

4 See the distinction between *logical* or *syntactic* and *metaphysical* types of *de re* necessity that Kit Fine draws and which have already been discussed in chapter 6, section 1, where it is stated that the interpretation offered in this book is intended as belonging only to the first type of *de re* necessity – the one interpreted by Kaplan in "Quantifying In." Footnote 3 of that chapter cites the reference to Fine's development of the distinction: Kit Fine, "The Problem of De Re Modality," in *Themes from Kaplan,* ed. J. Almog, J. Perry, and H. Wettstein (New York: Oxford University Press, 1989), pp. 197–202.

propositions and yet are propositions that are false. As a principle that collects only these true propositions, it will collect exactly the same propositions that express the *facts* that are referred to by the interpretation of real necessity; but it will collect these facts 'from above', as it were, since the principle neither refers to nor entails any particular contingent propositions that correspond to such facts. The advantage of collecting the propositions 'from above' is that it makes perspicuous the *a priori* character of the interpretation of real necessity that might escape our notice if the interpretation were explained in terms of facts alone.

[3] Kant's Epistemological and Epistemic Senses of Necessity. Given Kant's concern with the possibility of *a priori* knowledge, as described in the previous chapters, the necessity for him *is* limited to worlds we can know, and more particularly, to worlds whose forms are things we *can* know *a priori.* Consequently, Kant understands *a posteriori* necessity as an instance of *a priori* or *epistemic* necessity, just as nature and experience are for him instances of nature in general and experience in general (B164–65). Otherwise, Kant admonishes us, our claims of *a priori* knowledge would be applied erroneously to *things in themselves* – how things exist apart from sensibility (B164 *et passim*), an application that is actually impossible for us to bring about, he contends. And since it is impossible, *a posteriori* knowledge would also be impossible, since it is determined by *a priori* knowledge.

With regard to the specific epistemic modalities of Kant, he was convinced that if our knowledge is derived from experience – *a posteriori* knowledge – we cannot make any epistemic claims that can truly be said to be "strictly universal" (a universality that is not limited by our actual observations, past and present) or necessary, which, as previously noted, he ties to universality (B3–4). Since *a priori* knowledge is independent of experience, *a priori* knowledge alone has a chance of providing us with knowledge that is strictly universal and necessary, and hence extending over all possibilities in regard to our knowledge of objects that can be given to us. Kant therefore invokes the image of a "Copernican Revolution" in philosophy, in which knowledge would revolve around the mind rather than vice-versa, as is the case with realism. Such was the thinking in Kant's constructing an idealistic, or mind-centered, system of our knowledge of objects – what I have called his transcendental epistemology.

According to this system, we have seen, facts about real existences cannot affect our purely *a priori* ascriptions of necessary properties and relations to objects, since the properties and relations depend on just *a priori* facts about the mind, irrespective of any existing objects. For example, in regard to our representation of objects as outside us by means of outer sense Kant seems to just "discover" the fact about our mind that we represent objects as outside ourselves through Euclidean space – our form of outer intuition. Since our knowledge of this mental fact is independent of experience, it is *a priori*, thereby implying, as already indicated, that it is necessary and extends over all possibilities involving objects insofar as we can know them. A similar statement can be made about the mental fact that we can represent a succession of representations as objective only if we represent them through the concept of cause and effect. Of course, to be counted as more than a mere figment of the mind, the references of space and causality to existing objects, respectively, must be validated, and since the knowledge is claimed to be *a priori*, the validation itself must be *a priori*. The validation, we have found, consists in the *a priori* relation, or *a priori* reference, of the knowledge to empirical intuitions, and hence to appearances, and thus ultimately to initiators. Similar remarks can be made about the relation between time and inner sense and between the other categories and the 'I think'. We have traced the course of Kant's arguments to these effects in previous chapters.

Kant's *own* uses of *de re* necessity are as independent of experience as is the *a priori* knowledge whose possibility, or reference to objects, he wants to explain. But these uses of *de re* necessity belong to his (transcendental) *epistemology* of *a priori* knowledge – to his *theory* of such knowledge – and not to the knowledge that is under analysis in the *Critique.* Moreover, the epistemology is not on the same footing as the knowledge it is said to explain, since it contains only concepts – no sensible intuition constitutes the Fregean sense (*Sinn*) of any of his epistemological concepts. For example, his epistemological concept of a sensible intuition indeed has content belonging to a Fregean sense, but the sense is given only through general features belonging to the concept, and not though any intuition. These features have been given in chapter 3 above. Of course, the concept also has Fregean references (*Bedeutungen*), i.e. our human sensible intuitions. The latter are cases of sensible intuition. Kant's transcendental epistemology is thus, again, a system consisting exclusively of concepts. That is the truth of Strawson's assessment that Kant's total endeavor was, or in Strawson's opinion at least should have been, completely analytic in nature.[5] Only, in Kant's mind, the procedure of the *Critique* was synthetic, not analytic. Though they contain no intuitions, the tran-

5 P. F. Strawson, *op. cit.,* p. 16.

scendental concepts of the *Critique* are nonetheless related to one another in synthetic instead of analytic propositions, he maintains. The absence of intuition for Kant is no bar to acts of synthesis, provided nothing the equal of genuine knowledge (i. e. containing intuition as well as concept) is claimed for the ensuing propositions.

This distinction between the epistemology (the *explicans*) and the knowledge being explained (the *explicandum*) requires a corresponding distinction between two distinct uses of *de re* necessity in the *Critique.* The former (the epistemological) is one of the two proper subjects of the present book, whereas the latter (the epistemic) is not only the other subject of this book, but is, of course, the subject of the *Critique* as well. Again, the former is what has been called throughout "Kant's own uses of *de re* necessity," in his attempt to explain the latter, i. e. the uses of the necessity in our *a priori* knowledge of objects. And it is the latter alone that we have been talking about so far in this chapter as the object of Kripke's pronouncement that the necessity is *epistemic* and not *metaphysical.*

[4] The Abstraction of Kant's Idealism. In order to use our arguments from the last chapter as the basis for our realistic interpretation of the necessity, we must first abstract Kant's idealistic epistemology from the arguments. This can be done in a single stroke: the removal of *a priori* intuition. Chapter 3, section 4 argues that since for Kant the existence of objects can be given to a subject only through intuition, Kant's putative forms of our *a priori* sensible intuition – space and time – cannot belong to initiators, for otherwise they would belong to them without their affecting us. In that case, however, the existence of initiators *could* be given to us without their affecting us, since we could intuit the space and time in which they exist, and hence could intuit them through space and time independently of our *empirical* intuition of them, and thus independently of experience. But that would keep our intuition from being *sensible* and would instead make it *intellectual*; but for Kant that would make it impossible for the existence of initiators to be given to us in the first place, since, as sections 2 and 3 of chapter 3 explain, an intellectual intuition can belong only to the primordial being.

In addition, chapter 4, section 2 interprets a salient passage from the *Critique* as supporting a dichotomy between an intuition whose determination (say, space or time) *cannot* belong to initiators – our sensible *a priori* intuition – and one whose determination *can* belong to initiators (A26/B42). The latter are our *empirical* intuitions, the objects of which are not initiators *simpliciter*, that is, initiators referred to by Kant independently of our intuition of them, but rather our appearances of initiators. It is obvious that the elimination of *a priori* intuition eliminates the dichotomy, and with it, Kant's idealism.

[5] Kantian Derivatives. Two derivatives from our interpretation of Kant's account of epistemic necessity will appear in our interpretation of real necessity. The first is a condition that was introduced into our discussion in chapter 5, section 2. It was there argued that a proposition in which a singular term occurs in a subject position presupposes the existence of an object that is identified (à la Strawson) or reference to which is fixed (à la Kripke) by a description. This condition of identification through a presupposition of existence corresponds to the first premise of the derivation of Kant's *de re* necessity of Euclidean geometry that was put forward in the last chapter, viz., that we *intuit* things, or initiators, (as outside us), since for Kant *intuition is the representation of an object through which the existence of the object can be given to us*, and the object is identified through the terms in which its existence is given.

For example, if the proposition with respect to which *de re* necessity is in question is that water is H_2O, we can let the description 'the stuff from the tap' identify water and have the proposition presuppose the existence of the stuff from the tap. So, if the proposition in question is represented as *Qa*, where *Qx* stands for the property of being H_2O that is being ascribed to water and the identifying description is given a Russellian analysis, *Qa*'s presupposition of a proposition of the form $(\exists x)((Px \ \& \ (y)(Py \rightarrow y{=}x))$ and the latter's identification of or reference fixing to the object *a* as the stuff from the tap, which henceforth will be represented by the use of the iota operator as '$(\iota x)(Px)$', will be the first derivative from our interpretation of Kant's *de re* necessity of Euclidean geometry.[6]

The other derivative comes from Kant's proposition that we intuit things only through our *form* of intuition, a proposition that constituted premise 3 of our derivation of the necessity of geometry in the last chapter. The corresponding conceptual proposition in our interpretation of real necessity is that the existence of a given object of a proposition whose necessity is in question depends on its satisfying whatever condition is necessary for its existence. With regard to *Qa*, for example, an *instantiation* of the corresponding conceptual proposition in our realistic

6 See Manley Thomson, "Singular Terms and Intuitions in Kant's Epistemology," *The Review of Metaphysics* 26 (1972): 314–43 for an interpretation of Kant's 'intuition' as a representation that has many of the properties belonging to definite descriptions.

interpretation of necessity would be that the existence of the stuff from the tap depends on its being H_2O; that is, if the stuff from the tap exists, it is H_2O. Or let a reference to Bill Clinton be fixed as the most popular living president, whether former or present president. The same corresponding conceptual proposition would be instantiated by the proposition that the existence of the most popular living president depends on his paternity by William Jefferson Blythe Jr. It should be made it clear, however, that of course I am *not* saying that it is a conceptual truth that the existence of the stuff from the tap depends on its being H_2O, nor am I saying that it is a conceptual truth that the existence of the most popular living president depends on his paternity by William Jefferson Blythe Jr. The conceptual truth in both instances is only the *general* proposition that the existence of an object depends on its satisfying whatever condition is necessary for its existence, just as it is a truth in Kant's system that our outer intuition has a particular form, not that the system *of itself* requires that our outer intuition is spatial. That is not, nor does Kant claim it to be, such a truth in his system – indeed, he says he has no *a priori* reason that it must be so (B145) – even though he *does* claim that it is an *a priori* truth. He leaves it as a fact about the mind – just as he leaves the categories as the pure concepts of objects in general as facts about the understanding – that can be known independently of experience.

[6] The Initial and the Strengthened Arguments. Two arguments will now be given in support of the idea of real *de re* necessity, and the next section will give a more general argument for the same idea. The first argument will be called the *initial argument* and the second, the *strengthened argument.* (1)-(4) immediately below is the initial argument, and the strengthened argument, (1*)-(9*), follows it. Afterwards, a more general argument – called the *basic argument* – will be given in support of the idea.

The initial argument:

(1) Water is the stuff that comes from the tap.
(2) The stuff that comes from the tap doesn't exist unless it is H_2O.
(3) The stuff that comes from the tap exists.
Therefore,
(4) Water is H_2O.

As a logical consequence of (1)-(3) (4) is necessary, or the compound conditional proposition – in which (1)-(3) is the antecedent and (4) is the consequent – is necessary. But unless more is said in connection with

(1)-(3), it cannot be said that 4 is a *necessary* truth.[7] We would need more in connection with (1)-(3) to show that (4) is not, for instance, a *contingent* truth. That is, the validity of the argument would be unaffected if (4) were false; again, the conditional compound proposition would still be necessary, if the consequent, (4), were contingent. For instance, (2) might be false: We might live in a world in which we got something other than H_2O from the tap.

My idea of *real necessity* makes up for the gap between contingent and necessary truth by presuming it an *a priori* proposition that (4) presupposes (3), since presupposition is a logical relation between two propositions. Just as in chapter 9 the simple, single step of presupposition was all that was required to rebut the challenge to Kant's theory of necessity by Russell and Van Cleve, presupposition again is the simple, single step that is required here in support of the idea of real necessity. The strengthened argument contains this step and goes as follows.

(1*) (4) presupposes (3)	(*a priori*)
(2*) (2)	assumption
(3*) If (4) has a truth value, (4) is true	((from (1*) & (2*))
(4*) '(4) is false' is impossible	((by transposition of (3*)
(5*) (4) is not contingent	(from 4*)
(6*) "(4) and '(4) is contingent'" is impossible	(from (5*) and (4*))
(7*) If (4) is *merely* possible, "(4) and '(4) is contingent'" is possible	(definition of 'possible' and 'contingent')
(8*) (4) is not *merely* possible	(from (7*) and (6*))
(9*) (4) is not impossible	(from 4*)
Therefore,	
(10*) (4) is necessary	(from (9*), (8*) and (5*))

Whereas the initial argument does not justify the conclusion that (4) is necessary, the addition of presupposition to the initial argument, i.e.

7 A. N. Prior, *Formal Logic* (Oxford: Oxford at the Clarendon Press, 1962, second edition), pp. 210–11. Prior tells us that the Schoomen called the distinction I am drawing on here a distinction between *necessitas consequentis* (where the consequent is itself necessary – i.e. where the proposition is *de re* necessary) and necessitas consequentiae (where the consequent may not itself be necessary, but the necessity consists entirely in the rule that makes it a consequence of its antecendents – i.e. where the proposition is *de dicto* necessary).

the strengthened argument, does just that. Of course, in the strengthened argument, (2) is just as contingent as it is in the initial argument: It still is the case that we might live in a world in which the stuff from the tap isn't H_2O. But the validity of either argument doesn't require its necessity. The strengthened argument only demonstrates that *if* (2) is true, then (4) *must* be true (instead of (4)'s being merely true or contingently true, which is all that can be justified from the initial argument): In every possible world in which the fact expressed by (2) exists, or at every possible world at which (2) is true, (4) is true. The necessity of (4) thus consists in its truth at every possible world at which (2) is true. That is close to the desired conclusion that the initial argument does not reach.

It is close to the desired conclusion to account for the *de re* necessity of (4), but it isn't quite the conclusion we want, since we want the necessity of (4) to consists in its truth at every possible world at which *it has a truth value*, and not merely at every possible world *at which (2) is true*, even though if (4) has a truth value, (2) is true. What we want is the converse: If (2) is true, (4) has a truth value. Then, since we have so far established that (4) is true at every possible world at which (2) is true, we can conclude that (4) is true at every possible world at which it has a truth value, and we will get the conclusion that we desire, namely, that the necessity of (4) consists in its truth at every possible world at which it has a truth value. Its necessity could then be encapsulated as: If it has a truth value, it is true. Suffice it to say that the argument that follows our analysis of the strengthened argument – the basic argument – will give us the desired conclusion. But it will do so, not unexpectedly, only at a higher level of generality than we are prepared to go with the strengthened argument; that is, at a level of reasoning at which there is no reference to or entailment of any particular contingent proposition such as (2).

To get an argument that is *almost* as good as the one we would like and that gives us the conclusion that if (4) has a truth value, it is true, we need, as noted in the last paragraph, the *converse* of the proposition that if (4) has a truth value, (2) is true, which is all that we presently have. There is a simple method that will give us the converse, but it invites the possible objection that it only bypasses, without solving, the problem that keeps us from reaching our goal. The method is to introduce a *ceteris paribus* clause into the argument. We can maintain that if (2) is true, (4) has a truth value, if we hold that the truth of (2) is the *only* condition that is relevant to the argument and that determines that (4) has a truth value. One way of holding that it is the only such condition is to hold equal, or constant, whatever other conditions there might be regarding (4)'s having a truth value. This is the *ceteris paribus* clause: All other conditions of (4)'s having a truth value being equal, if (2) is true, (4) has a truth value. In other words, if (2) is true, (4) has a truth value, *ceteris paribus.*

As noted, we could then get an argument that is very close to the one we want, but we could do so only at the cost of using a *ceteris paribus* clause, with all its drawbacks. It would go like this: Since (4) is true at every possible world at which (2) is true, and (2) is true at every possible world at which (4) has a truth value, *ceteris paribus*, (4) is true at every possible world at which it has a truth value, *ceteris paribus.* Therefore, the necessity of (4) consists in its truth at every possible world at which it has a truth value, *ceteris paribus.* So, we get the conclusion we want, but only at the cost of using the *ceteris paribus* clause.

[7] Analysis of the Strengthened Argument (minus the *ceteris paribus* clause). The strengthened argument begins with the premise that propositions said to be realistically necessary presuppose the existence of the objects to which their subject terms refer; that is, it is argued that these realistically necessary propositions presuppose existential propositions to the effect that the objects of the necessary propositions exist. The relation of presupposition is employed in the sense introduced in chapter 5, viz., that the necessary propositions will have a truth value only if the presupposed existential propositions are true. The argument then goes on to *indirectly* relate the necessary propositions to certain further propositions that are *directly* and *truth-functionally* connected to the presupposed existential propositions. The relation is indirect, since it is mediated by the existential propositions: Given that the propositions to which *de re* necessity is to be ascribed, which will henceforth be called *the indicated propositions*, presuppose the existential propositions, and since the latter, but not the indicated propositions, are directly and truth-functionally connected to these further propositions, the indicated propositions are thus only indirectly related to these further propositions. The satisfaction of the presupposition of existence and the truth of the direct, truth-functional connections between the existence of the object(s) referred to and these further conditions together provide independent grounds for the ascription of real necessity to the indicated propositions.

The independence of these grounds for the ascription of real necessity to the indicated propositions is assured by the absence of presupposition from these grounds and its presence in the indicated propositions. Since the ascription of real necessity to the indicated propositions depends on *both* the presupposition and the grounds for the ascription, and since presupposition is missing from the grounds, real necessity cannot be ascribed to them. So, the absence of presupposition leads to the absence of real necessity from the grounds, which in turn gives a further reason for claiming that the grounds are independent from the indicated propositions, and thus can serve as genuine grounds for the ascription of real necessity to the indicated propositions. In sum, not only are the realistically necessary propositions only *indirectly* related to these further propositions that belong to the grounds for the ascription of necessity to the indicated propositions, but they are not *logically* related to them, either, since they are logically related only to the existential propositions – through presupposition – and the latter are only *truth-functionally*, but not logically, related to these further propositions.

The truth-functional connections between the existential propositions that are presupposed by the indicated propositions and the further propositions on which the real necessity of the indicated propositions is thereby based can be developed as follows. (The actual derivations of real necessities and the other realistically modal propositions will be given in the next chapter.) Let the indicated proposition to which real necessity is to be ascribed be represented as *Qa*. There is a logical form in which the presupposition of existence of the object $(\iota x)(Px)$ that is carried by *Qa* can be expressed. It is: $(\exists x)((Px \,\&\, (y)(Py \rightarrow y{=}x))$, where the condition *Px* is understood as identifying the object $(\iota x)(Px)$ in Strawson's sense of "identify,"[8] or "fixes the reference" to it, in Kripke's sense of "reference fixing," where a reference can be fixed through a *non-rigid designator* – that is, a referring expression that does not designate the same object in every possible world.[9]

Now that the *presupposition* condition on the real necessities has been specified as an existential proposition, a fact of a specific logical form or its logical equivalents can, in conjunction with the satisfaction of the presupposition condition, determine the real necessity of the presupposing proposition, *Qa*. If the condition *Qx* is a *materially necessary condition*

8 P. F. Strawson, *Individuals* (Methuen & Co: 1959), chapter 1.

9 The requirement that the term in question is one through which the *existence* of an object is presupposed corresponds to the similar condition on Kant's empirical intuition, and thus is determined by the interpretation of Kant's criterion of existence in general that is developed and given in chapter 3, section 3, above. For the argument that Kant's empirical intuition must indeed satisfy this condition see chapter 5, sections 4–7, above. For further specification of the manner in which an object is given see the immediately following paragraph in the body of the text.

of the existence of an object $(\iota x)(Px)$, where 'materially necessary condition' means *a consequent of a (truth-functional) material implication,* the fact that determines a real necessity is the satisfaction of a materially necessary condition by the object $(\iota x)(Px)$ whose existence is presupposed by Qa, if the fact is conjoined with the satisfaction of the presupposition condition. This fact can be expressed either as $(x)((\exists y)((Py \,\&\, (Px\rightarrow x{=}y))\rightarrow Qx))$, or as its logical equivalent. Qa is thus realistically necessary, if it presupposes the existence of the object $(\iota x)(Px)$ and the following conjunction is true: $\{(\exists x)((Px \,\&\, (y)(Py\rightarrow y{=}x))\} \,\&\, \{(x)((\exists y)((Py \,\&\, (Px\rightarrow x{=}y))\rightarrow Qx)\}$. That is, Qa is realistically necessary, given both the satisfaction of the presupposition of existence and the truth of propositions of the forms or their logical equivalents that flank the middle ampersand. Therefore, Qa is realistically necessary, given the satisfaction of the conditions of both the *presupposition* of the existence of the object $(\iota x)(Px)$ and the *existence* of the object itself, where the latter condition consists in the dependence of the existence on the condition Qx, i. e. consists in the proposition that the object satisfying the description $(\iota x)(Px)$ does not exist unless it is Q.

As already noted with respect to the initial argument, if presupposition is left out of the account, it is evident that only Qa, and not, *necessarily* Qa, follows (by *modus ponens*) from the mere conjunction of $(\exists x)((Px \,\&\, (y)(Py\rightarrow y{=}x))$ and $\{(x)((\exists y)((Py \,\&\, (Px\rightarrow x{=}y))\rightarrow Qx)\}$. In other words, the satisfaction of merely these conditions is not sufficient to give the result that Qa is necessary. For although these premises entail Qa, they allow that Qa may be false, and hence they fail to entail that it is necessary. Since presupposition is left out of the account, it allows that Qa may have a truth value in case Qx is *not* a condition of the existence of the object *a*. That is, it allows that Qa may be false, even though its truth follows from the truth of the premises. Only the truth of Qa, and not its necessary truth, follows from the premises, if they are not combined with presupposition, as least as far as the theory of real necessity is concerned.

On the other hand, the satisfaction of the presupposition condition alone, that is, apart from its combination with these same conditions, is also not sufficient to give the result that Qa is necessary. For if Qx is *not* a condition of the existence of the object $(\iota x)(Px)$, Qa could still presuppose the existence of the object. In that case, the object could exist, and yet Qa would *not* be necessary: It might be, say, contingent.

[8] 'Category Mistakes' and the Modalities. As a contradictory of a necessity, a contingency is logically incompatible with the same proposi-

tion's being necessary, whether the modalities are interpreted metaphysically, epistemically, or realistically. Their interdefinability entails this result. But the interdefinability must be determined in a uniform way; that is, the modalities are interdefinable only within a given interpretation of the modalities – metaphysical, epistemic, real, or what have you. Metaphysical contingencies, for example, should not be taken as contradicting real necessities – a prohibited cross-over that is crucial in the defense of the realistic interpretation of necessity if it is challenged by a metaphysical contingency.

Clearly, the falsity of *Qa* might be metaphysically possible even if *Qa* is realistically necessary. On the realistic interpretation of necessity, therefore, any attempt to make metaphysical contingency impossible would be futile. Rather, the only contingency the account of real necessity need exclude is *real* contingency – that, not metaphysical contingency, is the true logical contradictory of real necessity. The question, therefore, is whether or not the grounds adduced for the real necessity of *Qa* – $\{(x)((\exists y)((Py \mathbin{\&} (Px \rightarrow x=y)) \rightarrow Qx))\}$ plus the fact that *Qa* presupposes $\{(\exists x)((Px \mathbin{\&} (y)(Py \rightarrow y=x))\}$– allow *Qa* to be realistically contingent, and hence whether or not the grounds allow the *falsity* of *Qa* to be realistically possible. In other words, are the grounds sufficient to entail real necessity?

The entailment of real necessity, and thus the logical impossibility of real contingency, is the logical consequence of a chain of conditional propositions. If *Qa* presupposes the existence of the object $(\iota x)(Px)$, and if *Qx* is a materially necessary condition of the existence of the object, then if *Qa* has a truth value, *Qa* is true. Let us label this entire conditional '*A*' and its consequent (i. e. "if *Qa* has a truth value, *Qa* is true") '*B*'. As a conditional proposition that expresses a valid argument, *A* is a logical truth. The falsity of *B*, however, entails only that *Qx* is *not* a materially necessary condition of the existence of the object $(\iota x)(Px)$. It does not entail the disjunction that *either Qx* is not a materially necessary condition of the existence of the object $(\iota x)(Px)$ *or* the object doesn't exist. The disjunction is not entailed because the question of the existence of the object $(\iota x)(Px)$ cannot arise, since *ex hypothesi B* is false, and if it is false, *Qa* has a truth value; consequently, the object $(\iota x)(Px)$ exists, by presupposition.

So, since the entire conditional *A* is a logical truth that expresses a valid argument, *Qa* is realistically necessary iff *Qx is* a materially necessary condition of the existence of the object $(\iota x)(Px)$. Given the logical truth of *A*, a truth to which presupposition is essential, that *Qx is* a materially necessary condition of the existence of the object $(\iota x)(Px)$ is the fact that then makes *Qa* realistically necessary. Since real necessity is an interpretation of necessity that depends on the existence of a certain fact of a specific logical form or its logical equivalents (i. e. $(x)((\exists y)((Py \,\&\, (Px \rightarrow x{=}y)) \rightarrow Qx))$, the biconditional – that *Qa* is realistically necessary iff *Qx is* a materially necessary condition of the existence of the object $(\iota x)(Px)$ plus presupposition – identifies the fact that makes the necessity real, i. e. that Qx is a materially necessary condition of the existence of the object $(\iota x)(Px)$.

Now we turn to the impossibility that *Qa* is false. When that proposition is combined with the satisfaction of the presupposition condition, the falsity of *Qa* leads to a *reductio.* For, if *Qa* is false, it has a truth value. Since the existence of the fact that *Qx* is a materially necessary condition of the existence of the object $(\iota x)(Px)$, if combined with the presupposition of *Qa,* gives the result that if *Qa* has a truth value, it is true, then if *Qa* is false, it is true. This *reductio* shows that the falsity of *Qa* is logically impossible. This again leads to the conclusion that *Qa* is realistically necessary.

On the other hand, if *Qa* were a real contingency, it would have a fact of its own, with its own specific logical form or its equivalents, on which it would be based – a form to be given in the next chapter, but can here be abbreviated as the *negation* of the object $(\iota x)(Px)$'s dependence on *Qx* as a materially necessary condition of its existence. Since the real contingency of *Qa* would have a fact of its own on which it would be based, instead of refuting the realistic interpretation of *Qa*'s necessity, the realistic possibility of its falsity would actually count *in favor* of the realistic interpretation of *Qa*'s *contingency*, and thus indirectly confirm the realistic interpretation of the necessity. In other words, if the falsity of *Qa* were realistically possible, it would not count as a counter-example to the putative real necessity of the proposition, but would rather count as a case of real *contingency.* The alleged but misconceived counter-example is like a category mistake: The possibility of the falsity of *Qa* belongs to a different 'category', as it were, of real modality from necessity: It belongs to real contingency. Cases in one modality ('category'), while they can, and in certain logical relations, must, be logically contrary or contradictory to propositions in another modality cannot possibly count against the very legitimacy of another modality under a uniform interpretation of the modalities, if the realistic differences among the modalities are to be sustained and remain significant.

In this manner, the interpretation of real necessity is immune to any such alleged counter-example to its interpretation of necessity as the realistic possible falsity of *Qa*. The immunity comes from the logical interdefinability of the real modalities, a general logical property it shares with every interpretation of the modalities. In particular, real contingency is the negation of real necessity. To sum up, instead of constituting a counter-example to the realistic interpretation of necessity, the realistic possible falsity of *Qa* is an instance of real contingency, which supports, rather than undermines, the realistic interpretation of both of these modalities, since real necessity and real contingency are duly logically related to each other in the standard way in modal logic as logical contradictories – without the latter constituting a counter-example to the former. Far from constituting a counter-example, the latter occupies its rightful place in a modal 'square of opposition.'[10]

What then does real necessity reside in, if not the existence of certain facts of a specific logical form or its equivalents? Again, these would be facts in which *Qx* is a materially necessary condition of the existence of objects such as $(\iota x)(Px)$. Obviously, as already noted, it can't be these facts alone that provide for the necessity, since the propositions that express them yield as a consequence only the proposition that *Qa* is true. When the satisfaction of the presupposition condition is combined with them, however, we get propositions in which the property of having a truth value is ascribed to propositions such as *Qa*. It is the use of these propositions that leads to the employment of the concept of real necessity. So, though presupposition alone does not provide for the use of the concept of real necessity, or, for that matter, the use of the concepts of real possibility, impossibility, or contingency, and though the facts on which these uses depend do not by themselves provide for the uses, either, the combination of the facts with presupposition do provide for them. Were it otherwise, what is being offered here would simply not be an interpretation of necessity, or of the other modalities, either, since an interpretation (the *explicans*) should be logically independent of that which is being interpreted (the *explicandum*).

This analysis of the strengthened argument must not neglect the *a priori* determination of real necessity that was explained in section 2 of this chapter. That real necessity depends on the existence of metaphysically contingent facts needs to be complemented by what was earlier called the *principle of real necessity*. This is the trivial, but nonetheless for our purposes consequential, truth that the referent of a proposition does not exist unless it satisfies the necessary conditions of its existence. The

10 See A. N. Prior, *Formal Logic*, p. 187.

principle collects into a class all and only those propositions that correspond to the facts on which real necessities are based. This was the 'top down' way of determining real necessity that was given in section 2, and it needs to be part of an analysis of the strengthened argument such as is undertaken here, since it provides a way of understanding the idea of real necessity that is free of any reference to or entailment of any particular contingent fact. It thus provides a completely *a priori* exposition of the idea of real necessity.

[9] The Basic Argument. For some of us, philosophy is at its best when it tries to make its theories completely general, if theories are its objectives in the first place. A philosophical interpretation of *de re* necessity is no exception. So, if we can abstract from the strengthened argument a more basic argument, one whose premises do not refer or to or entail any particular contingent facts, which limit the range of possible worlds at which the real necessary truth of a given proposition is to be determined, or, alternatively, whose premises do not require that the *principle of real necessity* collect particular contingent propositions that determine the same limitation, or, finally, whose premises do not invoke a *ceteris paribus* clause, we shall have an argument all of whose premises are necessary, and, accordingly, a more general, and thus more basic, argument in support of the idea of real necessity.

Let p be any proposition in which a necessary condition of the existence of the referent of p is ascribed to the referent.

(1**) p presupposes the existence of its referent (*a priori*)
(2**) The referent of a proposition does not exist
unless it satisfies the necessary conditions of its existence (conceptual truth)
Therefore,
(3**) 'If p has a truth value, it is true' is necessary ((from (1**) and (2**))
(4**) If p has a truth value, it is a necessary truth ((from (3**))

[9] Comment on the Basic Argument. The *a priori* truth of (1**) consists in the logical relation that obtains between proposition p and its presupposition; and the conceptual truth of (2**) consists in the concept of 'necessary condition.' The necessity of 'If p has a truth value, it is true' obvi-

ously follows from the necessity of (1**) and of (2**). Finally, (4**), the necessary truth of p consists in the truth of p at every possible world at which p has a truth value.

As an ineluctable addendum, it must be said that this interpretation of real necessity immediately yields interpretations of real impossibility, possibility, and contingency. Impossibility is understood as the falsity of a proposition at every possible world at which it has a truth value, possibility as the truth of a proposition at a possible world at which it has a truth value, and contingency as possibility plus the falsity of the proposition at a possible world at which it has a truth value.

Clearly, the necessity of the proposition that if p has a truth value, it is true ((3**)) is more than the necessity that belongs to p merely as it occurs as the conclusion of a valid argument, which is all that it is in the initial argument. And the necessity of p itself in (4**) is more than the necessity of p that is partly derived from the truth of a contingent proposition such as (2), which is its force is in the strengthened argument. The necessity of p itself in (4**) is rather the conclusion of a valid argument all of whose premises are necessary. In freeing itself of particular contingent propositions such as (2), including propositions such as (2) that are collected by the *principle of real necessity*, and abstaining from using the *ceteris paribus* clause, the basic argument frees itself of particular contingencies that would otherwise effectively restrict its conclusions to just those worlds at which particular contingent propositions such as (2) are true. ((This is similar to Kant's attempt to free his *a priori* principles about the world from the particular judgments of the physical sciences that, being *empirical*, are contingent (cf. B165). He does so by citing alleged mental facts which he claims can be known *a priori*.))

The basic argument shows that when it is combined with a certain use of the notion of presupposition, the *principle of real necessity* of itself, that is, apart from its collection of a class of contingent propositions, has a certain implication. It is that a proposition does not have a truth value if it presupposes the existence of the object it purports to be about – its referent – and the object does not exist. Consequently, as the basic argument demonstrates, it is necessary that if a proposition *p* that ascribes such a condition to an object whose existence is presupposed by *p* has a truth value, *p* is true. As already noted, the basic argument provides for the encapsulation of the idea of real necessity: The real necessity of p thus consists in its truth at every possible world at which it has a truth value.

The basic argument is thus only a generalization from the strengthened argument. If we take as an *instance* of the basic argument – an instance that, unlike the basic argument itself, depends for its validity on a proposition that is a contingency – an argument about the proposition that water is H_2O, since the proposition presupposes the existence of the stuff that comes from the tap, and since that stuff doesn't exist unless it is H_2O, we can conclude that if the proposition that water is H_2O has a truth value, the proposition is true: its having a truth value implies its truth. In this example, instead of saying that being H_2O is *essential* to water, perhaps because of a Kripkean theory of rigid designation or more generally simply because one is necessary to the other's existence, we would rather say that the proposition that water is H_2O *has a truth value only if it is true*, and this is because of *both* the logical relation of presupposition *and* the fact that H_2O is a necessary condition of the existence of the stuff that comes from the tap. The common factor in both cases is that H_2O is a necessary condition of the existence of the stuff that comes from the tap, but the difference is that that seems enough for some, perhaps Kripke, to invoke essentialism, whereas essentialism can be avoided by invoking the theory of presupposition instead to interpret the necessity of the proposition that water is H_2O. And it has been our objective in this book to provide an account of *de re* necessity that studiously avoids such a commitment to essentialism.

[10] The Realistic *Contra* the Kantian Interpretation of *De Re* Necessity. The defense of this realistic interpretation of necessity is therefore no more than a reiteration of the point made at the beginning of the interpretation, viz., that like Kant's account of epistemic necessity, it is based on facts, and not merely propositions purporting to express facts. Alternatively, it is based on the trivial, yet nonetheless *a priori principle of real necessity*, that objects do not exist unless they satisfy the necessary conditions of their existence, where the principle determines exactly the same propositions that express the facts that belong to the correlative determination of real necessity. In the case of the realistic interpretation of the necessity, however, in contrast to Kant's epistemological account of it, the facts are taken to exist independently of the mind, that is, in reality, instead of existing in the mind alone and the alternative *a priori* determination of the relevant propositions is also determined independently of any mental facts.

If the abstraction of Kant's idealism from our interpretation of his explanation of the *de re* necessity of geometry and of the principle of causality is successful, we do not need his mental faculties to explain *de re realistically* necessary propositions about the existence of initiators in space and time; that is, unless we want to explain the possibility of our *a priori* knowledge of that fact. Even so, it is not clear that *every* explanation of that possibility must employ a mental apparatus, and *a fortiori*, an apparatus like Kant's. Nevertheless, since Kant's interest in explaining the possibility of our *a priori de re* knowledge of objects cannot without circularity be part of his explanation of that possibility, his considering himself compelled to employ the idea of mental faculties for that purpose at least tends to confirm the interpretive hypothesis that has been urged throughout, both in this book and in my previous book – that an explanation of the possibility of *a priori de re* knowledge is the major objective of the positive epistemological program of the *Critique.*

It should be clear that the qualification of our *a priori* knowledge as *de re* is essential to the remarks just made. Otherwise, the interpretation of Kant I have urged in both my previous book and in this book would be indistinguishable from Strawson's in a certain respect. As is well known, he considers such of Kant's propositions as that appearances can exist only in space and time as either very high *contingent* generalizations from experience[11] or as analytic propositions,[12] and hence as *a priori de dicto* propositions. In either case they would belong to the very framework within which we can experience the world. But since I am taking the relevant propositions of Kant's as necessary, and not contingent, and the interpretation of the *a priori* necessity as *de re*, not *de dicto*, I understand the propositions as neither contingent nor analytic. The attribution of necessity keeps them from being contingent, and the interpretation of the necessity as *de re* keeps them from being analytic.

On the other hand, since I am giving a *realistic* interpretation of *de re* necessity, and thus independently of the possibility of our *a priori* knowledge of objects, I have employed my interpretation of Kant's use of *de re* necessity by dispensing with his epistemic use of the necessity. Kant's implicit use of it can now be seen as just an instance of a more general rule.

Finally, as we have proceeded to follow Kant in our derivation of *real* necessity from its truth-functional grounds, our model has been our interpretation of Kant's derivation of the *de re* necessity with respect to appearances that belongs to geometry and causality given in sections 7 and 8 of the last chapter. A crucial step in the derivation of the necessity of geometry was step 3 of the derivation: "it is a conceptual truth for Kant that

11 Strawson, *The Bounds of Sense,* pp. 271–72.
12 Ibid., p 15 ff. *et passim*

if we intuit initiators as outside us, we do so through our form of outer intuition."[13] It is the form of our sensible intuition alone that entails that Euclidean geometry is necessary with respect to appearances alone and not with respect to things in themselves. Without step 3 of the derivation we wouldn't need the distinction between presupposition and entailment in our defense of Kant against Russell and Van Cleve. For if, *per impossible,* our form of sensible intuition *could* be true of initiators apart from their affecting us, as the so-called "neglected alternative" would have it, the Russell-Van Cleve attack on Kant would be otiose, since Kant would have already abandoned his theory that the mental faculties we in fact have alone explain the possibility of the necessary truth of Euclidean geometry in reference to appearances. A view of the reference of Euclidean space to initiators from the standpoint of what is being called a *realistic* interpretation of the modalities would concede at the outset the very point that Russell and Van Cleve are making against Kant. Their argument, after all, is that Kant's theory that the necessary truth of Euclidean geometry with respect to appearances is open to the objection that our cognitive constitution might be different from what it in fact is. If Kant were to adopt a *realistic* point of view and acknowledge that the truth of Euclidean geometry is actually independent of us, they would have no quarrel with him – and of course he would have no theory of Euclidean geometry – and no theory of the objectivity of the concept of causality, for that matter – that is quite his own, either.

[11] Other Accounts of Necessity. This presupposition is the same condition we find in the respective interpretations of the alethic modalities of Kripke and A. N. Prior. We have already seen that Kripke limits the possible worlds at which the ascription of a property or relation *Qx* to a given object rigidly designated as '*a*' is evaluated for truth at worlds in which the object *a* exists. Though in *Naming and Necessity* Kripke sidesteps the question of whether *Qa* has a truth value at worlds in which the object *a* does not exist, we have already attributed to him the view that the possible worlds at which the truth of *Qa* is to be evaluated are worlds in which the object *a does* exist.[14] This existence condition is iden-

13 It was premise 4 of the derivation of the ascription of the necessary properties and relations of Euclidean geometry to outer appearances in chapter 8 that identified the *a priori* form of our outer intuition as space.

14 I am given to understand that Scott Soames and Nathan Salmon have shown, independently of each other, that Kripke does not need to restrict the possible worlds at which a proposition about an object is to be evaluated for truth to

tical with the presupposition of existence in the account of *de re* necessity that I am drawing from my interpretation of Kant's uses of the necessity in the previous chapter.

The condition is also explicit in Prior's work on modal logic.[15] Prior maintains that the standard laws of equipollence between modal propositions – their interdefinability, already mentioned in section 7 above, fail in case the non-logical terms of the propositions are empty. In other words, the uses of the modalities are standard only if the presupposition of existence is satisfied. So, the presupposition of the existence of the object of a modal proposition is not distinctive to Kant's uses of the alethic modalities.

Besides the affinity that my idea of real necessity has to Prior's account of necessity, it departs in two major ways from Carnap's method for determining necessity. The first difference is that Carnap's basic term for necessity is "L-true," meaning that the truth of a sentence "can be established on the basis of the semantical rules" alone of a system that contains the sentence – independently of any extra-linguistic facts.[16] The idea of real necessity holds instead that the truth of a *de re* necessary proposition can be claimed to be established only on the basis of propositions purporting to express certain extra-linguistic facts, in agreement, of course, with the semantical rules of the language. Presupposition is the logical relation that connects the modal proposition to propositions about extra-linguistic facts. Whereas Carnap's necessary truth could be characterized by the common slogan "true in virtue of meaning alone," i.e. analytic truth, the idea of realistic necessary truth might instead be characterized as "true in virtue of reference (as well as meaning)," where reference involves the presupposition of an existential proposition about an object whose identification as ιPx[17] fixes the reference of the truth to an object, and where that existential proposition in turn materially implies that the identified object satisfies a certain further condition Qx.

---

worlds in which the object exists and that Kripke has agreed with their conclusion. But here I am simply going by what Kripke says in *Naming and Necessity.*

15 A. N. Prior, *Formal Logic,* pp. 188–89; see also his, "Modal Logic," *op. cit.*, p. 10.

16 Carnap, *op. cit.*, pp. 10 and 174. Though I am following Strawson and am speaking of the truth of *propositions*, Carnap speaks of the truth of *sentences.*

17 The iota operator signifies the uniqueness of an identification, which was otherwise previously represented contextually as $[(Px \,\&\, (y)(Py \rightarrow y{=}x)]$.

The second major difference between Carnap's approach and the one followed here is related to the first difference. Because Carnap's basic account of necessity is "L-true," substitution of identities in modal contexts is permissible only if the truths of the statements of identity, like the necessary truths themselves, are determined on the basis of semantical rules alone.[18] In other words, only expressions that denote identical intensions through semantical rules alone are substitutable for one another in modal contexts. As we saw in chapter 6, since Kaplan's method of quantifying into modal contexts works only for expressions that, like Carnap's expressions of intensions, denote abstract objects through "logical or . . . linguistic grounds alone," and thus can denote the objects necessarily,[19] Kaplan actually adopts Carnap's method in this respect.

But the method determining a real modality puts a different restriction on the possible substitution of co-referential terms in a modal sentence. Co-referential terms that can be substituted for a given referring expression in a sentence that expresses a modal proposition must be what will be called "type-identical" with the given expression.[20] For example, to take Quine's famous case, although the number of planets is now eight (given the demotion of Pluto), since the solar system does not determine any condition of the existence of a number, and since the realistic interpretation of necessary truths about numbers depends on necessary conditions of their existence, no description of a number in terms of the solar system can be substituted for a numeral in a sentence that expresses a necessary truth about the number.

So, in this respect, the interpretation of the modalities to be put forward here is closer to Kripke's interpretation (and even to Carnap's albeit cautious interpretation[21]) than to Carnap's (preferred interpretation) and Kaplan's interpretations, because the relation of presupposition that I am proposing logically connects the modalities to first-order extensional propositions about extra-linguistic facts, in contrast to Carnap's or Kaplan's respective semantical or linguistic rules that logically connect the modalities only to propositions about intensions or abstract objects, respectively.

---

18 Carnap, *op. cit.*, pp. 186–191.

19 Kaplan, *op. cit.*, p. 128.

20 The notion of a type-identical condition is explained in section 6 of the next chapter.

21 See footnote 18 above.

If we begin with real necessity, we can thus determine the truth-conditions of all alethically modal propositions about objects of first-order truth-functional logic, which allows our realistic modal propositions to be about more than Carnap's intensions and Kaplan's abstract objects, and hence whose truths depend on extra-linguistic facts instead of mere semantical or linguistic rules. Though Kripke would endorse the role of extra-linguistic facts in the determination of the truth of metaphysically necessary propositions, the method adopted here is also independent of the modal or counterfactual intuitive considerations that are basic to his interpretation of necessity in *Naming and Necessity*, in addition to its being independent of his theory of rigid designation.

[12] Prior and Kripke Style Semantics of the Modalities as *Transcendental*. The existential proposition satisfying the presupposition condition can also be said to be necessarily true in certain formal sense. If we adopt a Prior or a Kripke style semantics of the modalities, since a given object must exist in the possible worlds at which every proposition about the object is to be evaluated, the proposition that the object exists must be true at every such possible world, which makes it *necessarily* true. In this respect, both the existence of a given object and every necessary condition of its existence that it must satisfy receive the same treatment. As expressing the very limits on the possible worlds at which the truth of every proposition about the object is to be evaluated, these *limiting* propositions must be true at every such possible world, and hence necessarily true, even if only formally, simply because they themselves determine the possible worlds at which the truth of every proposition about the object is to be evaluated. This seems close to, if not identical, with the sense in which Kant called his Expositions of the Concepts of space and of time and his Deduction of the categories *transcendental*; only Kant considered the necessity of such truths to be *a priori* necessity, and hence, epistemically necessary. Consequently, it is *a priori* necessary for Kant that outer appearances are spatial, inner appearances, temporal, and spatial-temporal appearances, categorial.

[13] Independence from Transcendental Arguments. The propositions about a given object that we have just mentioned – that it exists and that it satisfy the necessary conditions of its existence depends on such and such conditions – are expressed in first-order logic without modality and their truth can be determined independently of the idea of truth at possible worlds, and *a fortiori* independently of truth at possible worlds delimited by the existence of the object and the necessary conditions of its existence. For example, the existence of a given object can be

expressed in a non-modal proposition having the form of the proposition $(\exists x)[(Px \,\&\, (y)(Py \rightarrow y=x)]$, where, in Kripke's respective senses of the following technical expressions, '$(\iota x)(Px)$' can be, but need not be, a "non-rigid designator" that "fixes the reference" of a "rigid designator" '*a*', which designates the same object in every possible world in which the object *a* exists.[22] And the dependence of that existence on a certain condition Qx can be expressed in a non-modal proposition having the form of the proposition $(x)[(\exists y)((Py \,\&\, (Px \rightarrow x=y)) \rightarrow Qx]$. The logic of the two propositions is *extensional*, whereas I am assuming that the logic of modal propositions is *intensional.*[23] Consequently, presupposition can relate propositions of intensional logic to those of extensional logic. This method of basing intensional propositions on extensional ones resembles Carnap's connection between his two "methods" of semantic analysis of sentences – extension and intension – and then applies both of them, albeit only very cautiously with respect to extensional sentences, to his semantics of the alethic modalities.[24]

[14] Summary. We can therefore take each of the two first-order non-modal propositions mentioned above – the presupposed proposition $(\exists x)[(Px \,\&\, (y)(Py \rightarrow y=x)]$ and the proposition expressing the necessary condition of the existence of the object, i.e. $(x)[(\exists y)((Py \,\&\, (Px \rightarrow x=y)) \rightarrow Qx]$ – and combine them by conjunction and determine the truth-conditions of the conjunction independently of modal or counterfactual intuitive considerations. This compound proposition will be the conjunction $(\exists x)[(Px \,\&\, (y)(Py \rightarrow y=x)] \,\&\, (x)[(\exists y)((Py \,\&\, (Px \rightarrow x=y)) \rightarrow Qx]$. It is a first-order extensional proposition whose form con-

---

22 See Kripke, *op. cit, passim.* The qualification "need not be" allows reference fixing by means of rigid designators, such as, for example, demonstratives, certain pronouns, and definite descriptions used in such a way that they are rigid.

23 See E.J. Lemmon, in collaboration with Dana Scott, ed. by Krister Sederberg, *An Introduction to Modal Logic* (Oxford: Basil Blackwell, 1977) p. 8, for an insistence that modal logic is "the study of intensional as opposed to extensional contexts." But the views just expressed in the body of text are actually closer to Carnap's, which take extensional and intensional forms of expression as more complementary than opposed to each other. See Carnap, *op. cit.*, p. 193 ff. *et passim.* Nonetheless, my interpretation of necessity is distinct from Carnap's, since mine is an interpretation of *de re* necessity, whereas his is explicated in the first instance in terms of *L-true*, which he takes to be an explication of Leibniz's notion of necessity and Kant's notion of analytic truth, and hence a *de dicto* form of necessity, p.8. Also see pp. 173 ff.

24 Carnap, *op. cit.*, p. 23 ff. *et passim*, especially chapter 5, "On the Logic of Modalities," p. 189.

stitutes the basis of our realistic interpretation of the *de re* necessity of the proposition $Q\alpha$, if $Q\alpha$ presupposes $(\exists x)[(Px \;\&\; (y)(Py \rightarrow y=x)]$.

***

After providing a further discussion of real modalities that focuses on the *differences* among the facts that determine different modalities, in contrast to this chapter, which has focused on both the presupposition condition that is common to all the real modalities and the modality of real *de re* necessity in particular, the next chapter will derive in a somewhat more, but definitely not entirely formal, way the four real modalities from truth-functional propositions plus presupposition. First, it will derive real necessity from certain truth-functional propositions plus presupposition, and then the derivation of real impossibility will follow from a single change in the derivation of real necessity. Real possibility will be almost immediately derived from real impossibility as its negation, and similarly for contingency. Finally, both real possibility and real contingency will again be derived, but in a slightly more complex way, a way that will employ results that we previously got from the derivations of real necessity and real possibility, respectively.

Chapter 11 – Derivations of the Real Modalities

[1] Real Modal and Non-Modal Propositions. The difference between propositions that can receive real modal ascriptions and those that are not subject to such ascriptions, and thus can be the bases for the ascriptions, can be made more precise than the way in which they were left in the last chapter. In the course of doing so, we will present the derivations of the real modalities in a relatively more formal way than the method of argument that was followed there.

A review of the role of the truth-functional logical forms of propositions as are referred to here as determining the real necessity of a proposition represented as Qa includes, first, a logical form of the presupposition condition on Qa, viz., $(\exists x)((Px \,\&\, (y)(Py \rightarrow y{=}x))$. This form can be used to express the proposition that, in identifying or fixing the reference to the object a, the object with which a is identified, viz., $(\iota x)(Px)$, satisfies the existential condition that is presupposed by Qa, or, alternatively, it is identified *as* $(\iota x)(Px)$ without necessarily (uniquely) satisfying the condition Px. Second, the particular logical form of the proposition that, in conjunction with the presupposed existential proposition, determines the real necessity of Qa is the truth-functional material implication $(x)((\exists y)((Py \,\&\, (Px \rightarrow x{=}y)) \rightarrow Qx)$ or any of its logical equivalents. Logical forms determining the real modalities of impossibility, possibility, and contingency will be provided later. The conjunction of the presupposition condition on Qa and the materially necessary condition of the existence of the object $(\iota x)(Px)$ will thus have the form $[(\exists x)((Px \,\&\, (y)(Py \rightarrow y{=}x))] \,\&\, [(x)((\exists y)((Py \,\&\, (Px \rightarrow x{=}y)) \rightarrow Qx)]$. A proposition of this form says that there exists an object $(\iota x)(Px)$ whose existence depends on its having the property Q. It is the truth of this latter proposition that is the ground of our true ascription of real necessity to the proposition Qa, when it is conjoined with the presupposition condition on Qa. Hence, so conjoined, it is the ground for the real necessity of Qa.

It might be asked, in the vein of a question posed in section 6 of the previous chapter, how is it possible to wring necessity from a truth-functional truth of the form $(x)((\exists y)((Py \,\&\, (Px \rightarrow x{=}y)) \rightarrow Qx)$? The answer, already indicated in same section of the chapter, is that the truth alone obviously can't do the job. It can do it only in conjunction with the presupposition condition. It is the logical relation of presupposition, a relation that, being logical, in a wide sense of the term, contains a logical necessity, which, in conjunction with the truth-functional truth in question, i. e. $(x)((\exists y)((Py \,\&\, (Px \rightarrow x=y)) \rightarrow Qx)$, imparts real necessity to the proposition *Qa* that carries the presupposition. Real necessity is the conjunction of this logical necessity with a truth-functional fact of a certain form or its logical equivalents. It is this necessity that turns an object's, i. e. $(\iota x)(Px)$'s, satisfaction of a necessary condition of existence, $(x)((\exists y)((Py \,\&\, (Px \rightarrow x=y)) \rightarrow Qx)$, into a realistically necessary property (or relation) *Q* of an object *a*.

In this respect, real necessity is indeed like physical necessity: Where the latter is derived from a physical law, which contains the necessity with which empirical rules determine the necessities of the *physical* world, the former is derived from truth-functional truths (which might [but need not] also be empirical rules) that are conjoined with the presupposition condition, which, like physical laws, contains the required necessity, and that, as a conjunction, they together determine necessities of the real world. One difference between the two is that where the necessity belonging to physical laws stands in need of an interpretation, that belonging to presupposition stands in need of a defense, a defense of a relation (presupposition) that doesn't observe the principle of bivalence with respect to propositions that are nonetheless significant.

Given the above distinctions between real modal propositions and non-modal propositions that can serve as their bases, there remain the distinctions among the modalities themselves that are essential to their use. For example, in relation to propositions that are realistically necessary at just certain possible worlds, there are of course other propositions that are merely realistically possible at those worlds, others that are realistically contingent, and still others that are realistically impossible. So, all the modal distinctions get covered by the idea of real modality. The truth-functional logical form $(x)((\exists y)((Py \,\&\, (Px \rightarrow x=y)) \rightarrow Qx)$, or its logical equivalents, of the facts that exist in any given world only limit the range of possible worlds that are covered by the modal properties, but within those limitations uses of all the modalities are still available.

[2] A Preliminary to the Derivations of the Modalities. Real necessity can, as already indicated, be derived from the satisfaction of two conditions, or two true propositions, which are metaphysically contingent. The other real modalities will then be determined according to the usual derivation of one modality from another – the so-called modal laws of equipollence, or interdefinability.

Let us again represent the proposition to which real necessity is to be ascribed as *Qa*, where *a* designates a given object and *Q* designates a property that is being ascribed to the object. The ascription of *real* necessity to *Qa* will use the standard symbol for necessity, viz., the box, '□', supplemented with the subscript '$_R$' as the sign of the realistic interpretation of the necessity, and thus '$\Box_R Qa$'. The first condition to be satisfied, if *Qa* is to be realistically necessary, consists in our agreement with Prior *et al.* on a condition, if *any* modality is to be ascribed to a proposition. It is that the object of the ascription exists. For the *real* modalities, this is the condition of presupposition. *Qa's* presupposition condition will thus be represented as $(\exists x)((Px \mathbin{\&} (y)(Py \rightarrow y=x))$, where the presupposed existing object will be represented as $(\iota x)(Px)$.

[3] Two Uses of Descriptions. It should be noted, however, that it can be successfully argued that the satisfaction of *Qa's* presupposition condition can be fulfilled if the object is merely *identified as* $(\iota x)(Px)$ but is not *actually* $(\iota x)(Px)$. That is, it is enough for $(\iota x)(Px)$ to identify the object, if $(\iota x)(Px)$ picks out the object for those involved in the identification even if the object so picked out is *not* in fact $(\iota x)(Px)$. Put another way, the object need not satisfy the condition $((Px \mathbin{\&} (y)(Py \rightarrow y=x))$ in order to satisfy the presupposition condition: The description $(\iota x)(Px)$ might just be sufficient to identify the object for the purpose of satisfying the presupposition condition without the description's actually being true of the object. It might, for example, be related to the object through a perception of the object, even if the perception is not veridical with respect to the object. Non-veridical perceptions of objects sometimes can be sufficient to identify objects even if the objects don't actually have the properties or relations that are exhibited in the perceptions of them. Or, if an object's satisfaction of an existential presupposition is still demanded, let the object satisfy the description, 'the object that is identified *as* the so-and-so.' In that case, $(\iota x)(Px)$ will represent *that* description – a second-level description, as it were, i. e. the one about being identified by a given first-level description. The two levels of description have been fully discussed vis a vis Kant in sections 4 and 5 of chapter 6.

For example, Kant's intuitions identify initiators even though in Kant's system the spatial-temporal properties and relations that are exhibited in our intuitions, and hence in our perceptions of initiators, are not true of the initiators unless the initiators are relativized to our outer and inner senses as appearances of the initiators. And Keith Donnellan has made somewhat famous examples of presupposition satisfaction in which perceptions serve to identify objects, and identify them as existing, even though the objects identified don't satisfy various features of the perceptions.[1] Donnellan argues that we should distinguish a use of a description where this sort of identification can take place from a use in which it cannot take place, but which requires instead that the description be true of the object in question. Donnellan calls the first use of a description *referential*, and the second, *attributive*.

As just mentioned, Donnellan's distinction is applicable to our epistemological interpretation of Kant's ontology. The referential use can be applied to our intuitions *of initiators* and the attributive use can be applied to *appearances*, understood as referents, or objects, *of* our intuitions.[2] Kant speaks of initiators independently of our spatial-temporal determination of them as appearances, whereas he doesn't speak of appearances independently of sensibility, and hence, *by implication within Kant's system* involving inner and outer sense, independently of space and time. Therefore, with respect to *initiators*, our intuitions can be accorded a *referential* use, whereas they must be given an *attributive* use in regard to *appearances* of the initiators. We fully discussed this difference in sections 3 and 4 of chapter 6, but we did so there without benefit of Donnellan's distinction between the two uses of descriptions. In sum, Donnellan's distinction between these two uses of descriptions could therefore be used as an additional way to distinguish between Kant's *initiators* and his *appearances.*[3]

Donnellan's distinction, however, will not affect the derivations of the real modalities. For the derivations will be determined instead by the truth-functional propositional connectives – material implication, conjunction, and negation – and the positive and negative conditions, *Qx*

1 Keith Donnellan, "Reference and Definite Descriptions," *Philosophical Review* lxxv (1966), 281–304. Kripke considers Donnellan's examples to be cases of "speaker's reference" and not "semantic reference." See his, "Speaker's Reference and Semantic Reference," in Peter A. French, Theordore E. Uehling, Jr. and Howard K. Wettstein, eds., *Contemporary Perspectives in the Philosophy of Language* (Minneapolis: University of Minnesota Press, 1979), pp. 6–27.

2 For remarks on this distinction between these two uses of the preposition 'of', genitive (possessive) and accusative (direct object), see chapter 5, section 1.

3 See Manley Thomson, "Singular Terms and Intuitions in Kant's Epistemology," who interprets Kant's use of 'intuition' as a definite description, but who does so without recourse to Donnellan's distinction between referential and attributive uses of descriptions. He has no need for the distinction, however, since the only objects he considers are Kant's *appearances*, not having drawn the epistemologically based ontological distinction between initiators and appearances as I have here.

and $\sim Qx$, involved in the implication, that form the factual bases of the real modalities.

Indifference to the *referential* use of $(\iota x)(Px)$ might, however, invite a dispute with Kripke and his followers. They would hold that since Px is a necessary condition of the existence of the object itself, if the object were *referentially* described, it would take special explanation to account for the *metaphysical contingency* of $(x)((\exists y)((Py \,\&\, (Px \rightarrow x{=}y)) \rightarrow Qx)$ that belongs to our account of *real necessity*, where in our account the description that fixes the reference of '*a*' in Qa, i.e. $(\iota x)(Px)$, can be a *non-rigid* designator. That is, according to Kripke and his followers, it would be hard to understand that in one possible world Qx would be a necessary condition of the existence of a given object, but in another possible world it wouldn't be. If an object has trans-world identity, it would seem that the necessary conditions of its existence would also have such an identity. On the other hand, the *attributive* use of $(\iota x)(Px)$ makes it easy to understand the metaphysical contingency in an interpretation of necessity such as ours. Being a non-rigid designator, $(\iota x)(Px)$ can identify an object in one possible world in which Qx *is* a necessary condition of its existence and another object in another possible world in which Qx *isn't* such a condition.

Without getting into a metaphysical argument with Kripke and his followers about whether an identical object can satisfy different conditions of existence in different possible worlds, it can be said, Kripkean concerns notwithstanding, that $(x)((\exists y)((Py \,\&\, (Px \rightarrow x{=}y)) \rightarrow Qx)$ can remain metaphysically contingent even if the description $(\iota x)(Px)$ is used *referentially*. For the condition that identifies the given object, $((Py \,\&\, (Px \rightarrow x{=}y))$, need not be true of the object, that is, the object might be identified only as *being identified by* $((Py \,\&\, (Px \rightarrow x{=}y))$, and still *that* identification might qualify as a non-rigid designator, and thus keep the implication in question, i.e. $(x)((\exists y)((Py \,\&\, (Px \rightarrow x{=}y)) \rightarrow Qx)$, contingent, which is required for the realistic interpretation of necessity. In a word, *being identified as* $(\iota x)(Px)$, where $(\iota x)(Px)$ is not true of an object, can be as contingently true of the object as *being* $(\iota x)(Px)$ can be contingently true of it. So, there is no advantage to insisting on the attributive use of descriptions in order to keep intact the realistic interpretation of necessity.

The relative indifference of the realistic interpretation of the modalities to Donnellan's distinction will reappear when the interpretation blocks opacity of reference, in section 6 below.

[4] The Second Condition of the Derivation of a Real Modality. The second metaphysically contingent fact that must exist if the derivation of

a real modality is to be valid is a fact of a logical form that can identify the modality in question. For example, real necessity is valid if the object that satisfies the presupposition condition, i. e. $(\iota x)(Px)$, also satisfies a material implication such as $(x)((\exists y)((Py \ \& \ (Px \rightarrow x=y)) \rightarrow Qx)$ or its logical equivalents. In sum, the derivation of $\Box_R Qa$ consists of the conjunction of these two metaphysically contingent propositions, $(\exists x)((Px \ \& \ (y)(Py \rightarrow y=x))$ and $(x)((\exists y)((Py \ \& \ (Px \rightarrow x=y)) \rightarrow Qx)$. It might be said, with qualification of course, that we will be deriving a necessity (albeit only a *real* necessity) from two metaphysical contingencies plus presupposition – a *real* "must" from two *metaphysical* "ises," supplemented by presupposition. Consequently, the derivation cannot be considered a completely formal proof.[4]

[5] The Procedure of the Derivations of the Real Modalities. The procedure of the derivations of the real modalities will be as follows. First, we will derive real necessity. From a single change in that derivation we will then derive real impossibility. After that, we will first give an almost immediate derivation of real possibility in the standard way by simply negating real impossibility. From this simple derivation and from the negation of one of the two principal premises of the derivation of real impossibility we will give a somewhat more formal derivation of real possibility. Finally, like the derivation of real possibility, we will first give an almost immediate derivation of real contingency as the conjunction of real possibility and the negation of real necessity. From this simple derivation and from the negation of one of the two principal premises of the derivation of real necessity we will give a somewhat more formal derivation of real contingency.

Each of the four derivations will be distinguished by one of two truth-functional propositional connectives – material implication or conjunction – plus, in two such cases, negation of the condition Qx that belongs to the proposition Qa whose modality is in question. Either of the two connectives will connect the presupposed existential proposition and the proposition that ascribes the condition in question, Qx or $\sim Qx$, to the object whose existence is presupposed.

Four principles correspond to the uses of the two connectives and the ascription of a given condition Qx or its negation $\sim Qx$. Consequently,

---

4 Although we should not forget that a completely *a priori*, and hence necessary, derivation of the *idea* of real necessity, as a "top down" argument, was given in the last chapter, section 9, as the basic argument.

each of the derivations of the four modalities will be distinguished by such a principle of derivation.

Following the derivations, each principle, or alternatively, its corresponding truth-functional connective and condition, will serve as a constraint on the co-referential terms that can be substituted for one another in the sentences that express their respective modal propositions. The principles, or the corresponding respective connectives and conditions, will thus function the same way that Kaplan's necessary denotation operator functioned for his interpretation of necessity in chapter 6 and that Kant's principles of sensibility and the understanding functioned in our interpretation of his account of epistemic necessity in chapter 8: These constraints block substitutions of co-referential expressions that turn true modal propositions into false ones. In other words, they protect the principles of the substitution of identities *salva veritate* and existential generalization, and thus block referential opacity from being induced by real modalities.

[6] The Derivation of the Real Necessity. Let *Qa* be the proposition that ascribes to gold, "as we have it," as Kripke says,[5] the atomic number 79. Let *Qa* presuppose the proposition that there exists a metal that possesses the properties we ordinarily recognize in gold, such as its malleability, ductility, glistening effect, etc. – properties that fix the reference of the term 'gold' to the object gold. Symbolically represent this presupposed existential proposition as $(\exists x)[(Px \,\&\, (y)(Py \rightarrow y{=}x)]$. Allow that the atomic number 79 is a necessary condition of the existence of the object we ordinarily recognize as being a metal that is malleable, ductile, glistening, etc. Represent this proposition symbolically as the material implication $(x)\ (\exists y)((Py \,\&\, (Px \rightarrow x{=}y)) \rightarrow Qx)$. The proposition that the existence of a metal of this description depends on its atomic number of 79 will instantiate the conceptual truth that the existence of an object depends on its satisfying the conditions that are necessary for its existence – a truth introduced in section 5 of the previous chapter as the *principle of real necessity.* We will call an *instantiation* of this principle, as it is in the present case of the atomic number of gold, the *principle of existential dependence.* Of course, this principle of instantiation does not entail that $(x)\ (\exists y)((Py \,\&\, (Px \rightarrow x{=}y)) \rightarrow Qx)$ is a conceptual truth, which of course it isn't. Employ the standard symbols for the truth-functional propositional connectives, identity, and quantifiers, let 'TV_{Qa}' symbolize that the particular proposition in question, viz. that gold has the atomic number

5 *Naming and Necessity,* p. 125

79, possesses the property of having a truth value; let '~' symbolize negation, '→' material implication, '⇒' entailment, and '⇛' presupposition. Let the derivation employ both *logical* and *real* modalities. The logical conforms to the principle of non-contradiction, which includes both meaning and conceptual relations as well as logical rules of valid inference. Let '◇' represent logical possibility, and '□', logical necessity. *Real* modalities limit logical modalities in the manner described in section 9 of chapter 10. Let '$\Diamond_R$' represent *real* possibility and '$\Box_R$', *real* necessity, where $\Box_R p$, if $\Box(TV_p \rightarrow p)$. The derivation of $\Box_R Qa$ goes as follows:

(1) $Qa \Rrightarrow (\exists x)((Px \ \& \ (y)(Py \rightarrow y{=}x))$	(presupposition)
(2) $(x)((\exists y)((Py \ \& \ (Px \rightarrow x{=}y)) \rightarrow Qx)$	(principle of existential dependence)[6]
(3) $TV_{Qa} \rightarrow Qa$ is true [7]	(from 1 and 2)
(4) Qa is false $\Rightarrow TV_{Qa}$	(definition of 'TV_{Qa}')
(5) Qa is false $\rightarrow Qa$ is true	(from 3 and 4)
(6) $\sim\Diamond(Qa$ is false)	(from 5)
(7) $\Box(TV_{Qa} \rightarrow Qa$ is true)	(from 6)
(8) $\Box_R Qa$	(from 7)

Observations. The conclusion can be expressed as: *Qa* is realistically necessary, since it is true at every possible world at which it has a truth value. Although we have only derived a necessary truth about gold, the *form* of the derivation is quite general and so is intended to be applicable to any set of suitably similar propositions about objects, their existence, and whatever conditions they satisfy, if they exist. Now that real necessity is derived, the other modalities can be immediately derived from it through the standard rules of modal interdefinability. Moreover, as already noted in section 9 of the previous chapter, a proposition p is realistically *impossible*, if it is false at every possible world at which it has a truth value; realistically *possible*, if it is true at a possible world at which it has a truth

6 I would like to thank Alan Berger for catching a technical problem with a previous symbolization of the principle of existential dependence.

7 I say '*Qa* is true' and not simply '*Qa*', since a proposition of the latter form may not have a truth value, whereas the derivation proceeds on the grounds that it does have one. The complication is due to the fact that our use of the very relation of presupposition requires that we allow that a proposition of the form *Qa* may not in fact have a truth value. Having been made explicit in the first two derivations, the complication can be ignored in the two remaining derivations – possibility and contingency.

value; and realistically *contingent*, if it is *both* realistically possible *and* it is false at a possible world at which it has a truth value.

Although the implication in 8 in the derivation is *materially* equivalent to Kripke's formula of metaphysical necessity, $Qa \rightarrow \Box Qa$, there are differences in how we reach our respective understandings of necessity. Whereas we arrive at ours on the basis of the notion of presupposition and, according to the principle of existential dependence, a case of an instantiation of the *principle of real necessity* (i.e. the conceptual truth that an object satisfies the necessary conditions of its existence), without reference to any intuitive modal or counterfactual considerations pertaining to gold, Kripke arrives at his in *Naming and Necessity* precisely on the basis of such considerations, plus his notions of rigid designation and truth at every possible world in which a given object exists.

The final observation concerns the connection between a second-level ascription of necessity to a proposition and first-level propositions about extra-linguistic facts. The proposition that gold has the atomic number 79 presupposes the proposition that a certain malleable, ductile, glistening etc. metal exists, which materially implies that the metal so described has the atomic number 79. But the conjunction of the two propositions – the presupposition of the proposition about the metal's existence – $(\exists x)[(Px \;\&\; (y)(Py \rightarrow y{=}x)]$ – and the material implication in which its existence is said to depend on its atomic number – $(x)[(\exists y)((Py \;\&\; (Px \rightarrow x{=}y)) \rightarrow Qx]$ – entails that if the indicated proposition that gold has the atomic number 79, i.e. Qa, has a truth value, it *must* be true. Despite the fact that both the presupposed proposition about the existence of the metal and that expressing its existential dependence on its atomic number are truth-functional first-order propositions about extra-linguistic facts, the modal proposition – a non-truth-functional second-order proposition about Qa, viz., that it is a necessary truth – can nonetheless be derived from the lower level propositions plus presupposition because it is logically related to the existential proposition (and to it alone) by way of presupposition and the latter materially implies that the metal has the atomic number 79.

[7] The Derivation of Real Impossibility. Not only can all the modalities be derived from one another in the standard ways by means of the modal laws of equipollence, but real impossibility can also be derived in a form that is similar to the one just employed in the case of real necessity.

We can derive real impossibility according to the form of argument we used for real necessity by making a single change in our derivation of the necessity: Simply take the condition substituted for Qx in Qa to be the *negation* of a necessary condition of the existence of a given object $(\iota x)(Px)$,[8] i. e. let $\sim Qx$ be a necessary condition of the existence of the object $(\iota x)(Px)$. Accordingly, let *Qa* be the proposition that Roger Clinton was Bill Clinton's biological father, and let *Qa* presuppose the proposition that in Bill Clinton's time there existed the owner of the Buick dealership in Hope, AR, where that description fixes the reference of 'Roger Clinton' to Roger Clinton. Symbolically represent that existential proposition as $(\exists x)((Px \,\&\, (y)(Py \rightarrow y{=}x))$. Since we are now deriving the real *impossibility* of the proposition *Qa*, a condition that cannot be true of Roger Clinton must be a negation of a necessary condition of his existence, i. e. $\sim Qx$. Therefore, let the existence of Roger Clinton depend on the *negative* condition that he was *not* the biological father of Bill Clinton. This proposition is a case of what I will call the principle of existential exclusion (since by negation it excludes a certain property as a condition of the existence of an object) and it will be represented symbolically as $(x)((\exists y)((Py \,\&\, (Px \rightarrow x{=}y)) \rightarrow \sim Qx)$. Lastly, employ the notations that were used in the preceding derivation of real necessity.

(1) $Qa \Rightarrow (\exists x)((Px \,\&\, (y)(Py \rightarrow y{=}x))$	(presupposition)
(2) $(x)((\exists y)((Py \,\&\, (Px \rightarrow x{=}y)) \rightarrow \sim Qx)$	(principle of existential exclusion)
(3) $TV_{Qa} \rightarrow Qa$ is false	(from 1 and 2)
(4) Qa is true $\Rightarrow TV_{Qa}$	(definition of 'TV_{Qa}')
(5) (Qa is true $\rightarrow Qa$ is false)	(from 3 and 4)
(6) $\sim\Diamond$(Qa is true)	(from 5)
(7) $\sim\Diamond$($TV_{Qa} \rightarrow Qa$ is true)	(from 6)
(8) $\sim\Diamond_R Qa$	(from 7)

On this interpretation of real impossibility, if the mere possession of a truth value – its being *either* true or false – by a proposition materially implies that the proposition is *false*, the proposition realistically *cannot* be true. Since the mere possession of a truth value by the proposition that Roger Clinton was Bill Clinton's biological father materially implies that the proposition is false, it is realistically impossible that it is true. The real impossibility is derived not from Kripke's "necessity of origin," regarding Bill Clinton's biological paternity,[9] but from the presupposition

8 Let the iota operator signify a definite description or an equivalent device that fixes the reference of the term '*a.*'.

9 Kripke, *ibid.*, p. 112.

of the indicated existential proposition and the latter's material implication of the indicated *negative* condition of the existence of Roger Clinton. The two impossibilities are not incompatible, of course. Further observations that are similar to those that were made about the derivation of real necessity can be made about this derivation of real impossibility as well. This is especially true with respect to the *formal* or *general* character of the derivation – that is, its applicability to any set of suitably similar propositions about objects, their existence, and whatever *negative* conditions are necessary for their existence.

[8] The Derivation of Real Possibility. Real possibility can be derived in the standard way from real impossibility by simply negating real impossibility. However, after concluding that it is realistically impossible that Roger Clinton was Bill Clinton's biological father, it would be confusing if we were to take the same proposition as our example of real possibility. So instead, for the purpose of deriving real possibility, let *Qa* be the proposition that Roger Clinton was Virginia Clinton's second husband, and let its presupposition be the same as before; that is, let the new proposition, like the original one, presuppose that during Bill Clinton's childhood there existed the owner of the Buick dealership in Hope, AR, and represent this existential proposition as $(\exists x)[(Px \;\&\; (y)(Py \rightarrow (y{=}x)]$. Again let that proposition fix the reference of the name 'Roger Clinton'.

If we accept the derivation of real impossibility, and if we accept that its negation, i. e. real possibility, contains presupposition, just as the rest of the real modalities contain it, then the negation entails the negation of just one of the two principal premises of the derivation of real impossibility, viz., premise 2, $(x)[(\exists y)((Py \;\&\; (Px \rightarrow x{=}y)) \rightarrow {\sim}Qx]$. So, as the negation of real impossibility, a principal premise of the derivation of real possibility is the proposition ${\sim}\{(x)[(\exists y)((Py \;\&\; (Px \rightarrow x{=}y)) \rightarrow {\sim}Qx]\}$. That is, it is the proposition ${\sim}(x)[(\exists y)((Py \;\&\; (Px \rightarrow x{=}y)) \;\&\; Qx]$. In terms of our example, the principal premise in question of the real possibility that Roger Clinton was Virginia Clinton's second husband is that during Bill Clinton's childhood there existed the owner of the Buick dealership in Hope, AR, *and* he was Virginia Clinton's second husband.

The gist of the reasoning is that the real possibility of *Qa* depends on, first, the presupposition of the existence of the owner of a Buick dealership in Hope, AR during President Clinton's childhood (*Px*) or, alternatively, of someone's being identified as such, and, second, the proposition that the owner was Virginia Clinton's second husband (*Qx*). In a word, it is the *co-existence* of these two conditions, Px and Qx, in the same object plus the presupposition of the satisfaction of the first condition that constitute the grounds for the real possibility that Roger Clinton was Virginia Clinton's second husband, or $\Diamond_R Qa$. Let us name the principle of such co-existence eponymously as the *principle of co-existence.* It is the principle that warrants the negation of the impossibility that Roger Clinton was Virginia Clinton's second husband, i.e. it warrants the negation of $\sim\Diamond_R Qa$, in case Roger Clinton was Virginia Clinton's second husband.

We now turn to the aforementioned "more formal" derivation of $\Diamond_R Qa$.

(1) $((\exists x)((Px \,\&(Py \rightarrow y{=}x)) \,\&\, Qx)$	(principle of co-existence)
(2) $\sim((x)((\exists y)((Py \,\&(Px \rightarrow x{=}y)) \rightarrow \sim Qx))$	(from 1)
(3) $\sim\,\sim\Diamond_R Qa$	(from 2 and from 1 and 8 of the derivation of real impossibility, *mutatis mutandis*)
(4) $\Diamond_R Qa$	(from 3)

[9] The Derivation of Real Contingency. Let the vocabulary be the same as above in regard to the real possibility that Roger Clinton was Virginia Clinton's second husband. Since real contingency is the conjunction of the real possibility of a proposition Qa and the *separate* real possibility of the proposition $\sim Qa$ – that Roger Clinton was not Virginia Clinton's second husband – we need two separate derivations of the respective accounts of the two real possibilities, one for $\Diamond_R Qa$ and the other for $\Diamond_R \sim Qa$. Otherwise, a contradiction would ensue, since in that case Qa and $\sim Qa$ would be taken to be true *together* of the same object $(\iota x)(Px)$ – that Roger Clinton both was and was not Virginia Clinton's second husband. Since real contingency consists in the conjunction of $\Diamond_R Qa$ and $\Diamond_R \sim Qa$, and since $\Diamond_R \sim Qa$ is the negation of $\Box_R Qa$, the grounds of real contingency would include the negation of the principle of existential dependence, which I will therefore call the principle of *existential independence.* In the present instance, instantiation of the latter principle follows from the negation of the proposition that Roger Clinton was Virginia Clinton's second husband that is described as realistically necessary *only for the sake of argument.* Whereas we previously derived real possibility by demonstrating the logical consequence of the negation of real impossibility, we can similarly derive real contingency by demonstrating the logical consequence of the negation of real necessity and then conjoining

it with real possibility. In other words, we will follow the method we used for deriving real possibility, only now the method will be applied to our previous derivation of real necessity, after which the result will be conjoined with real possibility.

The following argument is the more formal derivation of $\diamond_R{\sim}Qa$.

(1) $((\exists x)((Px \,\&(Py\rightarrow y=x)) \;\&\; {\sim}Qx)$	(principle of existential independence)
(2) ${\sim}((x)((\exists y)((Py \,\&(Px\rightarrow x=y))\rightarrow Qx))$	(from 1)
(3) ${\sim}\,\Box_R Qa$	(from 2 and from 1 and 8 of the derivation of real possibility, *mutatis mutandis*)
(4) $\diamond_R{\sim}Qa$	(from 3)

The conjunction of the two conclusions of the derivations of $\diamond_R Qa$ and $\diamond_R{\sim}Qa$, respectively, constitutes the derivation of real contingency: $\diamond_R Qa \;\&\; \diamond_R{\sim}Qa$.

[10] The Four Existential Principles as Restrictions on Substitutions of Identities. From the start of this book it has been explicit that one of its major objectives is to find a Kantian response to Quine's skepticism about *de re* modality, in particular, *de re* necessity. To review, according to Quine, modality induces opacity, which means that the principles of the substitution of identities and existential generalization are inoperative in opaque contexts. By blocking the operation of these principles, modal contexts are said to be not purely transparent. An intelligible interpretation of the objects of such contexts is therefore unavailable. The only way out, Quine maintains, is to restrict reference to objects to certain expressions: Only preferred expressions will be substitutable for identities and allow existential generalization. But the preferences entail, according to Quine, the adoption of the tradition associated with Aristotelian essentialism. For Quine and many other philosophers – most of his readers at the time that he wrote – it goes without saying that this is unacceptable.

Continuing the review, Kaplan contests Quine's skepticism regarding *de re* modality. Whereas Quine is right, that quantifying into modal contexts entails preferred expressions of reference, the latter does not entail Aristotelian essentialism. Kaplan follows Carnap when he argues that linguistic grounds can warrant the denotation of abstract objects by just certain expressions. These alone allow quantification into modal contexts without the adoption of Aristotelian essentialism. There is agreement, nonetheless, between Quine, on the one hand, and Carnap and Kaplan, on the other, that only a preference for just certain expressions, however that selection is made, will allow quantification into modal contexts. Only such a preference will allow observance of the principles of the substitution of identities and existential generalization.

The intention now, however, is not to challenge that condition on the observance of the two principles, and hence on quantifying into modal contexts. Consequently, *real* modalities must provide an alternative method for distinguishing between the eligible and the ineligible terms for possible substitution of identities and for possible existential generalization. It must thus eschew such distinctions as Kripke's between rigid and non-rigid designators, Carnap' designations of extensions and intensions, and Kaplan's distinction between necessary and contingent denotations of objects. This is not to say that those distinctions cannot or do not do the jobs they are meant for; it is only to claim that real modalities must have their own way of making the necessary distinctions between co-referential expressions that can and those that cannot be substituted for a given referential expression to a given object and from which existential generalization remains valid.

The method for making the necessary distinctions to keep the principles of substitutivity and existential generalization intact resides in the four existential principles or, alternatively, in the truth-functional propositional connectives – material implication, conjunction, and negation – and the positive and negative conditions, Qx and $\sim Qx$, that provided the basis for the realistic interpretation of the modalities in the preceding sections of this chapter. Each principle or corresponding propositional connective and condition determines uses of expressions that can be substituted for a given use of a referential expression in a use of a sentence that expresses a real modality. To make the point negatively, a use of an expression that is co-referential with a given use of a referential expression in the use of sentence that expresses a real modality is excluded as a possible substitution of the given use if it does not conform to the principle or corresponding propositional connective and condition that determines the real modality of the corresponding proposition. Accordingly, a use of a referential expression is substitutable for a given use of a co-referential expression in a use of a sentence that expresses a real modal proposition, if the use is determined by the same principle of existence or corresponding connective and condition that determines the real modality of the proposition that is expressed. This is what was meant by a "type-identical" condition in section 7 of the last chapter. Therefore, it is not the type of *expression* that determines which use of a co-referential expression(s) for a given object can be substituted for a given use of an expression that refers to the object, but the *existential principle* or its corresponding propositional connective and condition that makes the determination. As long as uses of expressions for an object are co-referential and the same condition *Qx* is ascribed to the object no matter how it is expressed, any expression will preserve substitution of identities and existential generalization, *provided the existential principle or its corresponding propositional connective and condition remains the same.*

For example, in accord with the principle of existential *dependence*, arithmetic conditions of the existence of numbers determine the uses of expressions for quantifying into the realistic necessary proposition that nine is greater than five, represented as $\Box_R Qa$. Contrary to Quine (but not the Quine of the *relational sense* of "propositional attitude"[10]), Kaplan, and Kripke, however, those same conditions would *not* exclude a use of the co-referential description 'the number of planets' as a possible substitution for a use of the numeral 'nine' or for a use of a variable referring to the number in a sentence that expresses $\Box_R Qa$, even if the use of the description fixes the reference of the numeral, since it does not matter that the solar system does not determine any necessary condition of the existence of the number. The only thing that matters in the determination of substitution of uses of expressions for identical objects is that the substitution be a use that is determined by the number's satisfaction of a necessary condition of its existence, which in the present instance is its being greater than five. Given the existential dependence on its being greater than five (*Qx*), since *Qa* presupposes the existence of the number, *Qa* must be true. The same principle would of course allow a use of 'three times three' to be substituted for a use of 'nine' or for a use of a variable to refer to the number, since the same principle of existence or its corresponding propositional connective and condition would be applicable.

On the other hand, if the operative principle is that of existential *independence* and its corresponding propositional connective is conjunction plus the negative condition that the object is *not* greater than five, then the use of 'the number of planets' would *not* be substitutable for the use of 'nine' in the sentential expression of $\Box_R Qa$. And it is this principle of existential independence that, on the realistic interpretation of the modalities, brings Quine, Kaplan, Kripke, as well as ourselves to reject the

10 W. V. Quine, "Quantifiers and Propositional Attitudes," reprinted in *Reference and Modality*, ed. Leonard Linsky, p. 101; original publication, *Journal of Philosophy*, vol. 53, 1956,

'the number of planets' as a possible substitution for the use of the numeral 'nine' in the use of the sentence that is expressive of the realistically necessary proposition in the present context. Again, it is not the expression itself that determines whether it is a possible substitution instance of a given co-referential expression or a variable for an object, or not, but rather a particular existential principle and its corresponding propositional connective and condition that determines a use of an expression or of a variable. The same expression, 'the number of planets,' can be the subject-expression of either sentence: 'the number of planets is necessarily greater than five' or 'the number of planets is (contingently) greater than five.' It all depends on the existential principle and its corresponding propositional connective and condition whether a use of 'the number of planets' is substitutable for the use of the numeral 'nine' in the use of the sentence or not.

As a further illustration, the principle of existential *exclusion* allows a use of 'the owner of the Buick dealership in Hope, AR' to be substituted for a use of the name 'Roger Clinton' in a use of a sentence that expresses the modal proposition that Roger Clinton could not have been Bill Clinton's biological father, even though the use of the description fixes the reference of the use of the name. Although not being Bill Clinton's biological father was a necessary condition of the existence of the owner, etc., i.e. Roger Clinton, it is not *impossible* that the owner, etc. was Bill Clinton's biological father, i.e. William Jefferson Blythe Jr. Since his being called 'Daddy' by Bill Clinton is not a condition of Roger Clinton's existence that excludes Bill Clinton's paternity, it does not conform to the principle of existential exclusion. On the other hand, the principle allows the use of 'Bill Clinton's adoptive father' as a substitution for the use of the name 'Roger Clinton' in the same sentence expressing the modal proposition. For the principle determines a class of negative conditions of Bill Clinton's paternity, and thus excludes uses of referential expressions that identify him otherwise.

Regardless of the principle that determines possible substitutions of identities and existential generalizations, whether belonging to Quine, Kaplan, Kripke, or the realistic interpretation of the modalities, however, Kit Fine's distinction between the logical question of *de re* modality and the metaphysical question of essentialism explains our limited use of such a restriction on substitution of identities. The self-imposed limitation to the logical question, adopted in chapter 6, keeps the present discussion from making any claim that purely arithmetic propositions about numbers express facts that are *essential* to them, whereas facts about the solar system are merely *accidental*. For example, the principle of existential *independence* allows substitution of uses of the same description, i.e. "the only man Bill Clinton called 'Daddy'" for 'Roger Clinton' in a use of a sentence that expresses the proposition that Roger Clinton took a hand to Bill Clinton's mother. But there is no speculation about either the necessity of origin being an instance of essentialism or Roger Clinton's physical abuse of Bill Clinton's mother being a condition that was inessential to his identity.

[11] Division of the Modalities. Regardless of the interpretation of the modalities, they can be divided into two groups, depending on whether a particular modality does or does not assign a truth value to a proposition at the actual world. Necessary and impossible propositions assign 'true' and 'false,' respectively, whereas neither possibility nor contingency assigns such a truth value. If a proposition is necessary, it must be true, and hence is true at the actual world, and if it is impossible, it must be false, and hence is false at the actual world, but if it is possible or contingent, its truth value at the actual world is undetermined.

This division of the modalities in general is reflected in the derivations of the real modalities that have just been completed. We derived necessity and impossibility from their respective material implications plus presupposition, but we followed the converse procedure and derived certain conjunctions plus presupposition from their respective antecedent possibilities and contingencies. How should we understand the latter derivations so that they don't suggest that a proposition is materially implied by its own possibility or contingency, that is, so that they don't suggest that a truth value of a realistic possible or contingent proposition is thereby determined at the actual world? The simple answer is that a real possibility or a real contingency, like other interpretations of possibility and contingency, is a truth only at a possible world, which does not entail that it also holds for the actual world.

In regard to this same question, the realistic interpretation of the modalities can be formulated in Kripke style semantics. Since the possibility or contingency of p holds only at worlds in which the subject of p exists (which, as already noted in the previous chapter, can be reformulated in my terms as p's presupposition of the existence of the subject of p, even though we have seen, also in the last chapter, that Kripke is indecisive on this point of presupposition and the withholding of a truth value), the possibility or contingency of p is to be understood as the truth of p at such a world. In addition, the contingency of p is also to be understood as the falsehood of p at such a world. Therefore, the material implications of the co-existences in question on the realistic interpretation of possibility and contingency can be understood Kripke style as the truth or falsehood of a proposition (expressing such a co-existence) at a possible world. But the ascription of neither possibility nor contingency to the properties or relations in respect to the subject of p entails that p is true or false at the actual world. Returning now to the realistic interpretation, it remains the case that whereas both real necessity and real impossibility entail their respective assignments of a truth value – true for necessity, false for impossibility – at the actual world, neither possibility nor contingency carries the same entailment.

[12] Summary. Quine's concerns about the intelligibility of *de re* necessity have now been dealt with, both in regard to Kant's possible response to them and in regard to the realistic interpretation of the necessity, and the modalities more generally, offered here. The responses have been based on the work of David Kaplan and guided by Kit Fine's distinction between the logical and the metaphysical forms of the necessity, where the logical form has been the one adopted here. Though the investigation began with adapting Kaplan to Kant, it was found that the form of the interpretation of Kant's epistemological uses and his epistemology of our epistemic uses of the necessity could be used to develop a post-Kantian interpretation of the modalities generally – an interpretation that has been called the *realistic* interpretation of the modalities. In the respect in which modal propositions, as instances of propositions about *intensions*, can be logically related to propositions about *extensions* – propositions of truth-functional logic with identity and quantification – our investigation into the real modalities can be viewed as following the two methods of semantic analysis proposed by Carnap – the methods of intension and extension. But real modality departs from Carnap, and thus from Kaplan, who follows Carnap in this regard, when it comes to the objects to which modal propositions are restricted. Carnap and Kaplan confine them to intensions and abstract objects, respectively, to which the propositions in question are related by semantical or linguistic rules, whereas the realistic interpretation offered here employs presupposition to relate modal propositions to propositions expressing extra-linguistic facts about objects that are concrete as well as abstract, if the use of the idea of existence is allowed with respect to abstractions, such as numbers. Though modal propositions are the objects of the semantic property of having a truth value, their logical relation to extra-linguistic propositions that can be expressed in first-order truth-functional logic provides them with a domain of propositions that is beyond the self-imposed lim-

its of the methods of Carnap and Kaplan. In this latter respect, the interpretation of the modalities provided here is closer to Kripke's interpretation. Thus, in an ironic way, an interpretation that has been adapted from our somewhat formal interpretation of both Kant's own epistemological uses of *de re* necessity and his epistemology of our own epistemic uses of the necessity – indeed, an interpretation claimed to have been found to be embedded in our interpretation of both of those uses – has come closer to Kripke's interpretation of the necessity than to Kaplan's, despite the fact that it was Kaplan's that provided the original insight for the formal work on Kant and thus for the logically independent interpretation of real *de re* necessity, indeed, for the realistic interpretation of all the modalities, that has now concluded our work on *de re* modality.

Chapter 12 – Conclusion

[1] Necessary and Contingent Properties and Relations of Objects. Section 8, chapter 10, argued that real contingencies cannot be genuine counter-examples to real necessities, even though contingencies in general are the logical contradictories of necessities in general, and therefore so are real contingencies the logical contradictories of real necessities. How can they be the logical contradictories of real necessities and not be genuine counter-examples? That is our first question, as we conclude our study of the modalities.

The real necessity of *Qa* determines every possible world at which the real contingency *Ra* has a truth-value, and hence *Qa* is true at every possible world at which *Ra* is false, where *Rx* is a condition for $(\iota x)(Px)$ to satisfy and where $(\iota x)(Px)$ identifies the object *a*. Therefore, if *Ra* has a truth value, it logically cannot contradict *Qa*. To put it another way, given the real necessity of *Qa*, if *Ra* has a truth value, it does so at a world at which $(\iota x)(Px)$ satisfies the condition *Qx*. That is why the condition that $(\iota x)(Px)$ can satisfy in a contingency is *Rx* and not *Qx*. Real necessities constitute the framework within which – determine the possible worlds at which – a real contingency has a truth value, and therefore the framework within which a real contingency can be false. Consequently, any property *R* that can be ascribed to an object *a* in a proposition that is a real contingency cannot be the property *Q* in a proposition that is realistically necessary. And that is why the real contingency *Ra* cannot be a counter-example to the real necessity of *Qa*, even though real contingency is the logical contradictory of real necessity. With respect to a given object, the properties in real contingencies must be logically compatible with properties in real necessities, if the real contingencies are to have truth values. And one way to assure this logical compatibility is to have the real necessities determine every possibility at which a real contingency has a truth value. In sum, if *Qa* is a real necessity, *Qa* cannot be a real contingency, that is, it cannot be the case that $\Diamond_R{\sim}Qa$, even though $\Diamond_R{\sim}Qa$ is the logical contradictory of $\Box_R Qa$. Reality is obviously abstracted from the propositions, such that only their logical forms remain, and it is those forms that give us the logical contradictories. A realistic interpretation of the modalities, however, in interpreting the logic with reality, makes real contingencies logically compatible with real necessities.

An alternative to the above reasoning leads to the same conclusion. Let the real contingency of *Qa* be not just the logical contradictory of the putative real necessity of *Qa*, but a counter-example, as well. In that case, *Qa* would *not* be realistically necessary. The material implication on which it is based would be false. But then, if the *principle of real necessity* obtains, as we argued in chapter 10, section 2 it must, if the interpretation of real necessity is to accept what seems an analytic truth to the effect that an object must satisfy the necessary conditions of its existence, an instantiation of the principle of real necessity as the principle of existential dependence will determine a case in which *another* necessary condition of existence, say, *Sx*, is satisfied by the object $(\iota x)(Px)$. In that case, the form of the above argument, that $\Diamond_R{\sim}Qa$ is only a logical, but not a real counter-example to $\Box_R Qa$, can be used again: $\Diamond_R{\sim}Sa$ is only a logical, but not a real counter-example to $\Box_R Sa$. $\Diamond_R{\sim}Qa$ can now be a real possibility, but since it now doesn't contradict $\Box_R Qa$, it is not a counter-example to it. Consequently, there is no point in insisting that $\Diamond_R{\sim}Qa$, besides being a logical contradictory of $\Box_R Qa$, is also a counter-example of $\Box_R Qa$, since the role of *Qx* can be assumed by *Sx*, and we are back where we started. So, we might as well accept *Qx* as a necessary condition of the existence of $(\iota x)(Px)$ at the start and therefore recognize that $\Diamond_R{\sim}Qa$ cannot be a counter-example to $\Box_R Qa$, though it is the logical contradictory of it.

The principle behind the above reasoning, whichever method is followed, is the logical relation of contradiction itself that obtains between necessity in general and contingency in general, and thus between real necessity and real contingency: If *Qa* is necessary, *Qa* logically cannot be contingent simply because it is the logical contradictory of contingency. Therefore, no contingency can be a counter-example to a true necessity; if a proposition is truly contingent, the corresponding putatively necessary proposition is not truly necessary. Consequently, with regard to real necessity and real contingency, since the former determines the very possibilities at which the latter has a truth value, there must be a true real necessity – a genuine one – that determines real contingencies, and which therefore is realistically immune to being contradicted by the contingencies.

So, our initial question, how a contingency can refute a necessity, belies a confusion. It fails to take into account the crucial difference between uninterpreted and interpreted propositions. Uninterpreted, $\Box Qa$ is logically contradicted by $\Diamond\sim Qa$. But if $\Box Qa$ *is interpreted as true,* $\Diamond\sim Qa$ must be false, in which case it could not refute $\Box Qa$. Real necessity is an interpretation of necessity, as real contingency is an interpretation of contingency, just as other interpretations of the modalities have a similar result. Once a proposition within the scope of the logical necessity operator $\Box$ is interpreted, the very logically contradictory relation it has to a corresponding interpreted proposition within the scope of (part of) the logical contingency operator $\Diamond\sim$ actually eliminates the contingency as a counter-example to the necessity. The only way a contingency can be more than a merely logical counter-example to proposition expressed as a necessity is for the expression of the necessity to be false. But once modal propositions are given truth values, they are interpreted, and that would be where our realistic interpretation of the modalities would come in.

Kant recognizes the validity of the above reasoning, I would submit – after all, the necessity of empirical intuition consists in its providing an interpretation of otherwise uninterpreted judgments consisting exclusively of concepts and logical operators. Consequently, in his epistemology he distinguishes between necessary and contingent propositions by way of his distinction between pure and empirical judgments, and between the pure a priori and a posteriori judgments. Kant maintained that there is no objectivity without necessity; so that the only objectivity that contingent propositions can have is within a framework whose dimensions are necessary. That is why it's impossible for the contingent to be a counter-example to the necessary, if the contingent is to be objective. Of course, this is not to deny that the contingent is the logical contradictory of the necessary. On the other hand, Kant also maintained that necessary judgments cannot be demonstrated as referring to any existing objects unless they can be shown to refer to empirical, and hence contingent judgments. Finally, he also maintains that the demonstration of the reference must itself be necessary, if the possibility of *a priori* knowledge is to be demonstrated.

[2] The Lawful and the Lawless – Kant's Theories of Knowledge and Morality. If there is one thing that concerns Kant in all of his philosophy, whether in regard to knowledge, morality, or beauty, it is the challenge he perceives the lawless to pose to the lawful. Mere regularity or rule-governedness falls short of lawfulness because of its contingency: Law, not mere regularity, entails necessity. Hence, regularity alone does not warrant the ascription of objectivity to a proposition. Hume showed as much. Consequently, experience cannot confirm principles of knowledge unless

its order is necessary. For Kant, that entails that sensations produced in us by things in themselves, though in fact regular, must brought under the control of the *a priori* forms of the mind – space, time, and then the categories – if they are to be part of our references to objects instead of merely referring to our individual, subjective mental states, regular though they may in fact be. Unless they are determined by these *a priori* forms, sensations exist outside of knowledge (A320/B376–77). In addition, desire and inclination are constantly in conflict in us with what we know to be our duty, which must exert itself over them by capturing our will and bending it to purposes that reason alone can dictate. The pleasure we get from beautiful objects, however, cannot be constrained by rational principles. Here reason must stand aside as we look to nature to give us analogies by which we can understand what makes things beautiful.

Where reason does hold sway – in knowledge and morality – there is a neat division in its relation to appearances, that is, to objects of our empirical representations. In knowledge, reason brings appearances under its rule, and empirical objects must conform to space and time, and thus mathematics, on the one hand, and to the categories, and thus the principles of empirical knowledge, on the other. In morality, however, the opposite is true; there reason works its will by excluding appearances from any role they might play in determining the will. The very objects of the will, and not merely the means by which its objects whose sources lie in desire and inclination are attained, arise from reason alone.

[3] Knowledge Entails Constraint (i.e. *Critique*) of Reason by the Senses. That reason must rely on appearances in the realm of knowledge if its own claims are to be confirmed – that appearances alone provide the actual cases that validate the claims of reason – is the very proposition that makes our intuition of objects *sensible*. This we saw in chapter 3. Were, *per impossible*, our intuition intellectual, it would not be necessary for initiators to affect us for us to know of their existence. Our intellect alone, as the source of the manifold of our intuition, would attest to their existence. But this would at once destroy the distinction between human knowledge and morality just drawn in the preceding paragraph. The distinction between knowledge and morality entails for Kant the distinction between our finding ourselves in a world we never made[1] and one in which our mere consciousness of our own existence would be sufficient

1 See the same comment in chapter 2, section 5, where the observation is attributed to both C. S. Peirce and Manley Thompson.

to attest to the existence of the rest of the world. Only the primordial being has this power, Kant says. For us, the power to bring objects into existence lies exclusively in the moral realm. These are none other than the objects reason tells us our conduct must pursue.

Once reason must depend on the senses to provide for the existence of objects to be given to us, the problem immediately arises how the objects whose existence can be given to us *must* conform to our forms of intuition – space and time – and then, those that do conform – appearances – how they in turn *must* conform to the categories. That is the *transcendental problem*, according to Kant, as seen from the standpoint of the contingent, representing existence as it does, instantiating the necessary: How is it that the contingent, which for us characterizes existence, *must* instantiate the necessary? From the standpoint of the necessary, the same problem becomes: How is it that the necessary *must* determine the contingent or existence, if it has not been derived or abstracted from the contingent in the first place, but is independent of it, which Kant claims it must be, if *a priori* knowledge is to be possible with respect to existing objects?[2]

But the transcendental problem is nothing less than making the lawless, despite its regularity, which is merely contingent, lawful. It is Kant's concern to make reason sovereign over sensations whose order could otherwise be explained no better than merely naturalistically, in Kant's estimation at least. Without its instantiation in experience, and hence in sensations, reason hovers aloft, unanchored to existence, as Kant describes Plato's contemplation of it (A5/B8–9), (and sensations, though they arise in a quite regular order out of the affect that initiators have on us, are without any object – they lack intentionality, that is, reference to any object, despite the orderliness of their appearance[3]). The transcendental problem, therefore, is precisely how *a priori* forms of the mind must determine the contingent regularity that is given to us in the existence of initiators, and therefore how these forms have reference to reality. The mind's determination of existence alone can validate the principles of reason, and thus metaphysics. This is the problem we have addressed in

2 Perhaps there are anticipations of modal system S4 here. Kant's transcendental expositions and deductions of concepts might be viewed as revealing convictions that are precursors of those that today are behind arguments for S4.

3 Again, as in section 2 of chapter 3, see Rolf George, *op.cit.*, for to my mind a largely correct analysis, that apart from their being contained in an intuition sensations for Kant do not refer to any object. Consequently, intentionality is not connected to them apart from their role in an intuition.

this book: To explain how Kant at once conceives of existence, which is independent of the human mind, and yet how the mind brings it under its control. So, the problem of the book is so basic to the *Critique of Pure Reason* that understanding Kant's major work must be considered unfinished until this problem is put to rest.

[4] Idealism and Realism. By nature, we are all realists. So Kant's solution to the transcendental problem requires an effort on our part to receive even a modicum of sympathy from us. Perhaps the only way to bring this sympathy into play is to keep reminding ourselves that Kant did not intend to reduce physical objects (which in his system are mere appearances) to mental representations – which is precisely what his idealism involves despite disclaimers of benighted self-appointed guardians who tirelessly argue to the contrary – *except*, that is, for his being wedded to reason. Only by interpreting physical objects as mere appearances, that is, as mere representations in the human mind (but not the existence of the objects that they represent – initiators, as chapter 2 has tried to make clear), can reason exercise its dominance over them and make them submit to the order that reason claims to confer on metaphysics. As the champion of reason, Kant becomes the protector of the "Queen of all the sciences" (Aviii), who otherwise is under attack by both the skeptics and the dogmatists. It was for no other epistemological reason than to explain the possibility of *a priori* knowledge that Kant made the natural world mental: As an epistemologist, he could see no other way to bring existence under the control of knowledge.

If this approach to Kant's idealism is correct, can we revert to our natural realism and let metaphysics as Kant conceived it pass into history? Recently, we have witnessed a spate of metaphysical investigations that steer clear of Kant's idealism and are realistic in nature. Strawson's is one, and Kripke's another, to name just two. What I have tried to do in chapters10 and 11 is offer yet another realistic approach to understanding how properties and relations can be said to be necessary with respect to existing objects. Perhaps ironically, it is an approach that actually emerged from the interpretation of Kant's arguments to the same purpose that has been given in chapters 2 through 9. It became evident that Kant's transcendental arguments, and in particular, his Transcendental Exposition of Space and Transcendental Deduction of the category of causality, were instances of a form that was even more general than he himself may have realized. That is, it was sufficiently general that it could accommodate a realistic interpretation of our ascription of necessary properties and relations to objects. Although reason would have to give up its claim to *epistemic* sovereignty over the form of existence, we could still intelligibly engage in discourse that unabashedly considers certain properties and relations as necessary to their corresponding existing objects.

The only special logical relation that was necessary to this endeavor was presupposition. But, acknowledging its arguably important non-adherence to the principle of bivalence, that relation created no other particular difficulty with regard to our present objectives, since it was introduced rather innocuously in chapter 5 as a mainstay of such establishment figures as Frege and Strawson, and as I pointed out there, it was even employed by the unlikely figure of Russell. Presupposition was the key not only to understanding Kant's idealistic theory of how knowledge refers to existing objects, but to a realistic account, marshaled in chapters 10 and 11, of necessary truth that was independent of Kripke's, at least to the extent that it stayed clear of intuitive modal and counterfactual considerations. It was therefore an account that was free of both Kant's idealism and Kripke's arguments for metaphysical realism in *Naming and Necessity.*

Perhaps there is no great need for yet another realistic account of necessity. But it is something new and interesting, I think, that a form of argument that seems to work for Kant apparently works as well for a post-Kantian realistic understanding of *de re* necessity, and of the rest of the modalities as well. We might thus extend our sympathy to Kant's endeavor to protect his "Queen of the sciences" while at the same time return to our natural inclination to approach the world in a more realistic spirit.

References

Allison, Henry E. *Kant's Transcendental Idealism: An Interpretation and Defense.* New Haven: Yale University Press, 1983, revised and enlarged, 2004.

– "Transcendental Realism, Empirical Realism, and Transcendental Idealism," *Kantian Review* **11** (2006).

Bennett, Jonathan. *Kant's Analytic.* Cambridge, U.K.: Cambridge University Press, 1966.

Burge, Tyler. "Belief *De Re,*" *Journal of Philosophy* 77 (1974).

Carnap, Rudolph. *Meaning and Necessity: A Study in Semantics and Modal Logic.* Chicago: University of Chicago Press/Phoenix Books, 1947, supplemented 1956.

Donnellan, Keith. "Reference and Definite Descriptions," *Philosophical Review* **75** (1966).

Dummett, Michael. *Frege: Philosophy of Language.* London: Duckworth, 1973.

Evans, Gareth. *The Varieties of Reference,* ed. John McDowell. Oxford: Clarendon Press, 1982.

Falkenstein, Lorne, "Critique of *Kantian Humility,*" *Kantian Review,* **5** (2001).

Fine, Kit. "The Problem of De Re Modality," in eds., J. Almog, J. Perry, and H. Wettstein, *Themes from Kaplan.* New York: Oxford University Press, 1989.

Frege, G., "The Thought: A Logical Inquiry," translated by A.M. and Marcelle Quinton. *Mind,* **65** (1956), reprinted in ed., P. F. Strawson, *Philosophical Logic.* London: Oxford University Press, 1967.

– "On Sense and Reference," trans. Max Black, in eds. Peter Geach and Max Black, *Translations from the Philosophical Writings of Gottlob Frege.* Oxford: Basil Blackwell, 1952.

George, Rolf, "Kant's Sensationism," *Synthese* , **47** (1981).

Greenberg, Robert. *Kant's Theory of A Priori Knowledge.* University Park: Penn State Press, 2001.

Guyer, Paul. *Kant and the Claims of Knowledge.* New York: Cambridge University Press, 1987.

Hanna, Robert. *Kant and the Foundations of Analytic Philosophy.* Oxford: Clarendon/Oxford University Press, 2001.

Kant Immanuel. *Kritik der reinen Vernunft* (1781/1787) ed., Raymund Schmidt. Hamburg: Felix Meiner Verlag, 1956, reprinted 1990. The paginations of the first, 1781, and second, 1787, are cited as A and B, respectively. English translations employed: Norman Kemp Smith, London: Macmillan, 1929; Werner S. Pluhar, Indianopolis/Cambridge: Hackett, 1996; Paul Guyer and Allen Wood, Cambridge, UK: Cambridge University Press, 1998.

Kaplan, David. "Quantifying In," in ed., Leonard Linsky, *Reference and Modality.* London: Oxford University Press, 1971. Originally appeared in eds., D. Davidson and J. Hintikka, *Words and Objections: Essays on the Work of W. V. Quine.* Dordrecht-Holland: D. Reidel Publishing Co., 1969.

– "Opacity," in eds., Lewis Edwin Hahn and Paul Arthur Schilpp, *The Philosophy of W. V. Quine.* La Salle, IL: Open Court, 1986.

– "*Demonstratives*," in eds., Joseph Almog, John Perry, and Howard Wettsetein, *Themes from Kaplan.* New York, Oxford: Oxford University Press, 1989.

Kripke, Saul. *Naming and Necessity.* Cambridge, MA: Harvard University Press, 1980, originally published 1972.

– "Speaker's Reference and Semantic Reference," in Peter A. French, Theordore E. Uehling, Jr. and Howard K. Wettstein, eds., *Contemporary Perspectives in the Philosophy of Language.* Minneapolis: University of Minnesota Press, 1979.

Langton, Rae. *Kantian Humility: Our Ignorance of Things in Themselves.* Oxford: Clarendon Press, 1998.

Lemmon, E. J. (in collaboration with Dana Scott). *The Lemmon Notes: an introduction to modal logic*, ed. Krister Segerberg. In ed., Nicholas Rescher, *American Philosophical Quarterly Monograph* series, no. **11.** Oxford: Basil Blackwell, 1977.

McDowell, John. *Mind and World.* Cambridge, MA, London: Harvard University Press, 1996.

– *Meaning, Knowledge, and Reality.* Cambridge, MA: Harvard University Press, 1998.

Perry, John. "Frege on Demonstratives," *The Philosophical Review* **86** (1977).

Prauss, Gerold. *Kant und das Problem der Dinge an sich.* Bonn: Bouvier Verlag Herbert Grundmann, 1974.

Prior, Arthur N. *Formal Logic.* Oxford: Oxford at the Clarendon Press, 1962, second edition.

– "Logic, Modal" in ed., Paul Edwards, *The Encyclopedia of Philosophy*, vol. **5.** New York: Macmillan Publishing Co. & The Free Fress, 1967.

Quine, W.V., *From a Logical Point of View.* Cambridge, MA: Harvard University Press, 1980, originally published 1953.

– *Ontological Relativity and Other Essays.* New York: Columbia University Press, 1969.

– *Word and Object.* New York: The Technology Press of the Massachusetts Institute of Technology and John Wiley & Sons, Inc., 1960.

Rescher, Nicholas. "Kant and the 'Special Constitution' of Man's Mind: The Ultimately Factual Basis of the Necessity and Universality of *A Priori* Synthetic Truths in Kant's Critical Philosophy," in *Studies in Modality*, American Philosophical Quarterly Monograph Series, 8. Oxford: Oxford University Press, 1974.

Russell, Bertrand. "On Denoting," *Mind*, **14** (1905), reprinted in ed., R. C. Marsh, *Logic and Knowledge.* London: Allen & Unwin, 1956.

– *The Problems of Philosophy.* London: Oxford University Press, reprinted 1957 (first edition 1912).

– *Our Knowledge of the External World* (delivered in Boston as the Lowell Lectures) 1914, reprinted, New York, Mentor Books, 1960.

– "The Philosophy of Logical Atomism," *The Monist* (1918), reprinted in *Logic and Knowledge*.
– *An Inquiry into Meaning and Truth*. New York: W.W. Norton, 1940.
Smit, Houston. "Kant on Marks and the Immediacy of Intuition," *The Philosophical Review*, **109** (2000).
Smith, Norman Kemp. *A Commentary to Kant's 'Critique of Pure Reason'*. New York: The Humanities Press, 1918, reprinted 1950.
Smullyan, Arthur F. "Modality and Description," *The Journal of Symbolic Logic*, **13** (1948), reprinted in ed., Leonard Linsky, *Reference and Modality*. London: Oxford University Press, 1971.
Strawson, P. F. "On Referring," *Mind* **59** (1950), reprinted in ed., Antony Flew, *Essays in Conceptual Analysis*. London: Macmillan & Co. Ltd., 1956.
– *Introduction to Logical Theory* (New York: John Wiley & Sons, Inc., 1952.
– *Individuals: An Essay in Descriptive Metaphysics*. London: Methuen & Co., 1959.
– *The Bounds of Sense*. London: Methuen & Co., 1966.
– *Skepticism and Naturalism: Some Varieties*. New York: Columbia University Press, 1985.
– "Intellectual Autobiography of P. F. Strawson" in ed., Lewis Edwin Hahn, *The Philosophy of P. F. Strawson*,. Chicago & La Salle, IL: Open Court Publishing Co., 1998.
Thompson, Manley. "Singular Terms and Intuitions in Kant's Epistemology," *The Review of Metaphysics* **26** (1972).
– "Things in Themselves," in *Proceedings and Addresses of The American Philosophical Association*. Newark, DE: The American Philosophical Association, 1983.
Van Cleve, James. *Problems from Kant*. New York, Oxford: Oxford University Press, 1999.
Walker, Ralph C.S. *Kant, The Arguments of the Philosophers*. London: Routledge & Kegan Paul, 1978.
Wilson, Margaret D. "The 'Phenomenalisms' of Berkeley and Kant," in ed., Allen W. Wood, *Self and Nature in Kant's Philosophy*. Ithaca: Cornell University Press, 1984.
Wood, Allen, Guyer, Paul, and Allison, Henry E., "Debating Allison on Transcendental Idealism," *Kantian Review*, vol. 12–2, 2007.

Index

Bei Fragen zur Produktsicherheit wenden Sie sich bitte an:
If you have any questions regarding product safety,
please contact:

Walter de Gruyter GmbH
Genthiner Straße 13
10785 Berlin
productsafety@degruyterbrill.com